AF361906

Reconfigurable Antennas

Reconfigurable Antennas

Editors

Dimitris E. Anagnostou
Michael Chryssomallis
Sotirios K. Goudos

MDPI • Basel • Beijing • Wuhan • Barcelona • Belgrade • Manchester • Tokyo • Cluj • Tianjin

Editors

Dimitris E. Anagnostou
Institute of Signals, Sensors
and Systems
Heriot Watt University
Edinburgh
United Kingdom

Michael Chryssomallis
Electrical and Computer
Engineering Department
Democritus University
of Thrace
Xanthi
Greece

Sotirios K. Goudos
School of Physics
Aristotle University
of Thessaloniki
Thessaloniki
Greece

Editorial Office
MDPI
St. Alban-Anlage 66
4052 Basel, Switzerland

This is a reprint of articles from the Special Issue published online in the open access journal *Electronics* (ISSN 2079-9292) (available at: www.mdpi.com/journal/electronics/special_issues/reconfigurable_antennas_1).

For citation purposes, cite each article independently as indicated on the article page online and as indicated below:

LastName, A.A.; LastName, B.B.; LastName, C.C. Article Title. *Journal Name* **Year**, *Volume Number*, Page Range.

ISBN 978-3-0365-4282-9 (Hbk)
ISBN 978-3-0365-4281-2 (PDF)

Contents

About the Editors

Dimitris E. Anagnostou

Dimitris E. Anagnostou (S'98–M'05–SM'10) received his B.S.E.E. degree from the Democritus University of Thrace, Greece, in 2000, and the MSEE and Ph.D. degrees from the University of New Mexico, USA, in 2002 and 2005, respectively. From 2005 to 2006, he was a Postdoctoral Fellow with the Georgia Institute of Technology, USA. In 2007, he joined the SD School of Mines & Technology, USA, as an Assistant Professor, where he received a promotion and tenure. In 2016, he joined the Heriot Watt University, Institute of Signals, Sensors and Systems (ISSS), Edinburgh, UK. He has also worked at the Kirtland AFB, NM, in 2011 and at the Democritus Univ. of Thrace, Greece, as an Assistant Professor.

Dr. Anagnostou is currently supported by a prestigious H2020 Marie Skłodowska-Curie Individual Fellowship on wireless sensing technologies (ViSion-RF). He has authored or co-authored more than 160 publications (h-index 28; more than 3000 citations), two book chapters, and holds two U.S. patents. His interests include: antennas (reconfigurable, adaptive, miniaturized, space/satellite, and wearable), microwave circuits, radar sensing, phase-change materials (e.g. VO_2) for reconfigurable devices, direct-write, MEMS, and applications of artificial neural networks in electromagnetics and health care.

Dr. Anagnostou serves or has served as an Associate Editor for the IEEE Transactions on Antennas and Propagation, IEEE Access, and IET Microwaves, Antennas and Propagation. He is a Guest Editor for IEEE Antennas and Wireless Propagation Letters and MDPI. He is a member of the IEEE AP-S Educational Committee and Vice Chair for online content. He received the 2010 IEEE John D. Kraus Antenna Award, the 2011 DARPA Young Faculty Award, the 2014 Campus Star Award by the ASEE, the 2017 Young Alumni Award by the University of New Mexico, the H2020 MSCA Fellowship, and 4 Honoured Faculty Awards. His students have also been recognized with IEEE and university awards.

Michael Chryssomallis

Michael T. Chryssomallis received his Diploma and the Ph.D. degree in Electrical Engineering from Democritus University of Thrace (DUTh) and he is a Professor in Electrical and Computer Engineering (ECU) Department in DUTh. Over recent years, he has taught/teaches undergraduate and graduate courses in Antennas, Telecommunications, Mobile Communications, Wave Propagation, and Electromagnetic Interferences and Immunity (EMC/EMI). He has supervised more that 80 diploma theses, 14 postgraduate theses and 2 PhDs (+2 in progress). He has participated in 17 research projects. Regarding his administrative work, he was elected as the Sector Director for 5 years, Deputy of Department Chair for 4 years and Director of Postgraduate Studies for 4 years.

He worked with the Communications Group (director Prof. P. S. Hall) of the University of Birmingham for the period of October 1997 to January 1998, in the areas of Active Antennas, and with the Wireless Group (director Prof. C. G. Christodoulou) of the University of New Mexico for the periods of April to June 2000, and April to July 2002, in the areas of Microstrip Antennas and Arrays, and Smart Antennas.

His current research interests are in the areas of microstrip antennas, small antennas, RF-Mems, smart antennas, algorithms of beamforming and angle of arrival estimation, and propagation channel characterization. He is the author or co-author of about 100 journal papers and conference proceedings (>1400 citations in Google Scholar and >850 in Scopus). He has translated an English technical book into Greek and has co-authored three chapters in Wiley Encyclopedias of Electrical and Electronic Engineering and RF/Microwave Engineering. He serves as a reviewer for several publications of IEEE, IEE and others. He has been a member of Greek Technical Chamber and of IEEE since 1988 (senior member since 2000).

Sotirios K. Goudos

Sotirios K. Goudos is an Associate Professor in the Department of Physics, Aristotle University of Thessaloniki, Greece. Prof. Goudos is the director of the ELEDIA@AUTH lab member of the ELEDIA Research Center Network. He serves as an Associate Editor for IEEE Transactions on Antennas and Propagation, IEEE Open Journal of the Communication Society, and IEEE ACCESS.

Prof. Goudos is the founding Editor-in-Chief of the *Telecom* open access journal (MDPI publishing). He is also member of the Editorial board of the *Electronics* open access journal (Microwave and Wireless Communications Section). Prof. Goudos is currently serving as the IEEE Greece Section Secretary.

Preface to "Reconfigurable Antennas"

Antennas and metasurfaces that can operate in different complex environments will be part of every modern wireless communication network, such as 5G, Internet-of-Things (IoT), and radar sensing. These new networks require antennas and metasurfaces with a high degree of reconfigurability. To meet the performance requirements for each application scenario, reconfigurable antennas, advanced phased arrays with adaptive nulling, multiple beams, low sidelobes, reconfigurable metasurfaces, and different signal processing techniques provide effective solutions. Such kinds of antennas and metasurfaces are commonly used in several fields of applications, such as surveillance, tracking, and search and rescue. Of particular interest are novelties in the element design and materials, novel switching and performance-boosting technologies, system architecture, and the array radiation and scattering properties of novel metasurfaces. As these research areas have different developmental statuses and trends, this book examines the current state of the art and project future research directions of reconfigurable antennas and metasurfaces for solving different design problems. Despite substantial progress over the last twenty years, all these antenna devices and metasurfaces are still the subject of intense research aiming to reach their full potential.

This book is addressed to engineers and researchers on antennas and metasurfaces, globally, as well as to academic staff, faculty and graduate students, who are likely to find useful information in the review and technical articles. The authors are all experts in the fields of antennas and metasurfaces and have deep knowledge in the field.

In this book, 10 papers are published, covering various aspects of antenna technology, from antenna boosters to radio-frequency switching using novel materials.

Mohamadzade et al. [1] present recent developments and state-of-the-art flexible and conformal reconfigurable antennas. Flexible and conformal antennas rely on non-conventional materials and realization approaches, and thus, despite the mature knowledge available for rigid reconfigurable antennas, conventional reconfigurable techniques are not translated to a flexible domain in a straightforward manner. There are notable challenges associated with the integration of reconfiguration elements such as switches, the mechanical stability of the overall reconfigurable antenna, and the electronic robustness of the resulting devices when exposed to the folding of sustained bending operations. This review paper discusses various approaches to realize flexible reconfigurable antennas, categorizing them on the basis of reconfiguration attributes, i.e., frequency, pattern, polarization, or a combination of these characteristics. The challenges associated with the development and characterization of flexible and conformal reconfigurable antennas and the strengths and limitations of available methods are reviewed considering the progress made in recent years, and open challenges for future research are identified.

J. Anguera et al. [2] introduce antenna boosters. Antenna boosters are non-resonant small antenna elements that excite currents on the ground plane of a wireless device and do not rely on shaping complex geometric shapes to induce multiband behavior, but rather the design of a multiband matching network. Their design approach results in a simpler, easier, and faster method for creating multiband and efficient antennas. Since multiband operation is achieved through a matching network, frequency bands can be configured and optimized with a reconfigurable matching network. The authors present two kinds of reconfigurable multiband architectures with antenna boosters. The first one includes a digitally tunable capacitor, and the second one includes radiofrequency switches. The results show that antenna boosters with reconfigurable architectures feature multiband behavior with very small sizes, compared with other prior art techniques.

Iftikhar et al. [3] present a new model for radiofrequency switching technology called magnetostatic responsive structure (MRS) that enables reconfigurable operation in 4G/5G antenna frequency bands. Specific frequencies of interest include but are not limited to advanced wireless services (AWS) from 2.18 GHz–2.2 GHz, mid-bands of sub-6 GHz 5G (2.5 GHz and 3.5 GHz), and 4G bands around 600 MHz/700 MHz, 1.7 GHz/2.1 GHz/2.3 GHz/2.5 GHz. The ABCD matrices of the MRS switch are derived from the S-parameter values and are shown to be a good model from 100 kHz to 3.5 GHz. Agreement between simulations, analytical results, and circuit model values is shown.

Koutinos et al. [4] present a robust PIN diode-based reconfigurable antenna for WLAN/WiMAX applications. A super-shape elliptical radiator is used to achieve wider intrinsic bandwidth compared to the classical rectangular patch antenna, while the dimensions remain comparable. The proposed antenna is fed at two points, exciting both horizontal and vertical polarizations but at different operating frequencies. To achieve a wider bandwidth, as a whole but also for each polarization, the symmetrical feeding points for each excitation are also employed with a proper feeding network. PIN diodes are also used in the feeding network to provide the option of a narrower bandwidth.

Yang [5] presents a reconfigurable 3D slot antenna for 4G and sub-6G smartphones with a metallic casing. The antenna is located at the bottom of a smartphone and is integrated with the metallic casing. PIN diodes are loaded at the dual-open slot and the folded U-shaped slot, respectively, which realize four working states. The antenna has a compact volume and covers the long-term evolution (LTE) bands of 698–960 MHz and 1710–2690 MHz, and the sub-6G bands of 3300–3600 MHz and 4800–5000 MHz. The design is fabricated together with the measured performance and a comparison to other state-of-the-art antennas shows that the proposed design has multiband characteristics with a small size.

Shereen et al. [6] presents a novel combo-reconfigurable antenna architecture for the frequency and pattern reconfiguration of a novel antenna system for 5G millimeter-wave mobile communications. The tuning system independently controls the frequency and radiation pattern shifts, without letting them affect each other. The proposed antenna consists of two patches, radiating at 28 GHz and 38 GHz. A negative-channel metal–oxide semiconductor (NMOS) transistor is used as an ON/OFF switch. Frequency reconfiguration is controlled by two switches that vary the patch dimensions. Pattern reconfiguration is achieved by the remaining 16 that connect and disconnect parasitic stubs to the ground to control the surface currents. The results are shown for 18 different combinations of the states of the switches.

A compact reconfigurable UWB MIMO antenna with four radiators that accomplish on-demand band rejection from 4.9 to 6.3 GHz is presented by Khan et al. in [7]. Two radiators are then placed perpendicularly to each other to exploit the polarization diversity, leading to high miniaturization. Two additional radiators are then fixed obliquely on the same laminate (without increasing its total size) in an angular configuration at a ±45 ° angle to the first two planar radiators to further exploit polarization diversity. The design is validated by measurements of a fabricated prototype. On demand band rejection through the use of PIN diodes, wide impedance matching (2–12 GHz), high isolation amongst the radiators, compactness achieved by angular placement of the radiators, low gain variation over the entire bandwidth, band rejection control achieved by adjusting the gap between stub and ground plane, and low total active reflection coefficient (TARC) values make the proposed design very suitable for commercial handheld devices.

The optimal design of an aperiodic reconfigurable microstrip antenna array for broadcasting applications is introduced by Gravas et al. in [8]. The array is designed to serve as a DVB-T base station antenna operating in a single broadcasting channel (optimized at 698 MHz, center frequency of DVB-T channel 49). This allows the array to concurrently achieve a particular radiation pattern shaping with high forward gain, main lobe tilting, and null filling inside the service area, as well as low sidelobe level outside the service area. To concurrently satisfy all the above requirements, the geometry dimensions and the array feeding weights (amplitudes and phases) are optimized, thus leading to a complex multi-variable and multi-objective problem. The problem is solved by applying a recently developed particle swarm optimization (PSO)-improved variant, called PSO with velocity mutation (vm), in conjunction with the CST electromagnetic simulation software, which is employed by the PSOvm every time a full-wave analysis is required. Furthermore, all the optimization methods found in the CST environment are compared with the PSOvm. The results show that the PSOvm is capable of producing an antenna array geometry that is closer to the predefined requirements than the geometries derived by the rest of the optimizers in the least amount of computational time.

Alharbi et al. [9] discuss the application of metasurfaces for three different classes of antennas: reconfiguration of surface-wave antenna arrays, realization of high-gain polarization-reconfigurable leaky-wave antennas (LWAs), and performance enhancement of Van Atta retrodirective reflectors. The proposed surface-wave antenna is designed by embedding four square ring elements within a metasurface, which improves matching and enhances the gain when compared to conventional square-ring arrays. The design for linear polarization is comprised of a 1 × 4 arrangement of ring elements, with a 0.56λ spacing, placed amidst periodic patches. A 2 × 2 arrangement of ring elements is utilized for reconfiguration from linear to circular polarization, where a similar peak gain with better port isolation is realized. A prototype of the 2 × 2 array is fabricated and measured; a good agreement is observed between the simulations and the measurements. In addition, the concepts of the design of polarization-diverse holographic metasurface LWAs that form a pencil beam in the desired direction with a reconfigurable polarization are discussed. Recent developments incorporating polarization reconfigurability in metasurface LWAs are briefly reviewed. In the end, the theory of Van Atta arrays is outlined and their monostatic radar cross-section (RCS) is reviewed. A conventional retrodirective array is designed using aperture-coupled patch antennas with a microstrip-line feeding network, where the scattering from the structure itself degrades the performance of the reflector. This is followed by the integration of judiciously synthesized metasurfaces to reconfigure and improve the performance of retrodirective reflectarrays by removing the abovementioned undesired scattering from the structure.

Last but not least, Farooq et al. [10] present a highly compact frequency-selective surface (FSS) that has the potential to switch between the X-band (8 GHz–12 GHz) and C-band (4 GHz–8 GHz) for RF shielding applications. The proposed FSS is composed of a square conducting loop with inward-extended arms loaded with curved extensions. The symmetric geometry allows the RF shield to perform equally for transverse electric (TE), transverse magnetic (TM), and 45 ° polarizations. The unit cell has excellent angular stability up to 60 °. The resonance mechanism is investigated using the equivalent circuit models of the shield. The design of the unit element allows for the incorporation of PIN diodes between adjacent elements for switching to a lower C-band spectrum at 6.6 GHz. The biasing network is on the bottom layer of the substrate to avoid effects on the shielding performance. A PIN diode configuration for the switching operation is also proposed. In simulations, the PIN diode model is incorporated to observe the switchable operation. Two prototypes are fabricated, and the switchable operation is demonstrated by etching copper strips on one fabricated prototype between adjacent unit cells (in lieu of PIN diodes) as a proof of the design prototypes. Comparisons among the results confirm that the design offers high angular stability and excellent performance in both bands.

The editors would like to take this opportunity to thank all authors of the respective works for their high-quality contributions. It is our hope that readers will find new and useful information on antennas and related technology for 5G, IoT, and sensing applications. Moreover, the editors wish to express their thanks and acknowledgements to the anonymous expert reviewers who ensured the published works are of the highest quality.

This work was supported in part by the European H2020 Marie Skłodowska-Curie Individual Fellowship under grant # 840854 (VisionRF).

The editors wish to dedicate this book to:

Dimitris E. Anagnostou—to my mentors, my children, Ioanna, and my parents Manos and Mariki.

Michael Chryssomallis—to my family.

Sotirios K. Goudos—to my wife Athena and my daughters Mary and Mandy.

Dimitris E. Anagnostou, Michael Chryssomallis, and Sotirios K. Goudos

Editors

Review

Recent Developments and State of the Art in Flexible and Conformal Reconfigurable Antennas

Bahare Mohamadzade [1],*, Roy B. V. B. Simorangkir [1], Sasa Maric [1], Ali Lalbakhsh [1], Karu P. Esselle [2] and Raheel M. Hashmi [1]

[1] School of Engineering, Macquarie University, Sydney, NSW 2109, Australia; roy.simorangkir@ieee.org (R.B.V.B.S.); sasa.maric@mq.edu.au (S.M.); ali.lalbakhsh@mq.edu.au (A.L.); raheel.hashmi@mq.edu.au (R.M.H.)

[2] School of Electrical and Data Engineering, University of Technology Sydney, Sydney, NSW 2007, Australia; Karu.Esselle@uts.edu.au

* Correspondence: bahare.mohamadzade@students.mq.edu.au

Received: 14 July 2020; Accepted: 21 August 2020; Published: 25 August 2020

Abstract: Reconfigurable antennas have gained tremendous interest owing to their multifunctional capabilities while adhering to minimalistic space requirements in ever-shrinking electronics platforms and devices. A stark increase in demand for flexible and conformal antennas in modern and emerging unobtrusive and space-limited electronic systems has led to the development of the flexible and conformal reconfigurable antennas era. Flexible and conformal antennas rely on non-conventional materials and realization approaches, and thus, despite the mature knowledge available for rigid reconfigurable antennas, conventional reconfigurable techniques are not translated to a flexible domain in a straight forward manner. There are notable challenges associated with integration of reconfiguration elements such as switches, mechanical stability of the overall reconfigurable antenna, and the electronic robustness of the resulting devices when exposed to folding of sustained bending operations. This paper reviews various approaches demonstrated thus far, to realize flexible reconfigurable antennas, categorizing them on the basis of reconfiguration attributes, i.e., frequency, pattern, polarization, or a combination of these characteristics. The challenges associated with development and characterization of flexible and conformal reconfigurable antennas, the strengths and limitations of available methods are reviewed considering the progress in recent years, and open challenges for the future research are identified.

Keywords: flexible antennas; frequency reconfigurable; microfluidic antennas; pattern reconfigurable; polarization reconfigurable; reconfigurable antennas; wearable antennas

---

1. Introduction

Reconfigurable antennas have received a lot of attention in modern communication systems due to their versatile applications by offering extra functionality. A reconfigurable antenna is defined as a single structure that is capable of switching between one and a combination of characteristics such as frequency, pattern, and polarization [1]. These structures with their flexible and multi-operation characteristics offer compact and cost-effective antennas for modern communication systems such as cellular radio systems, satellite communication, airplane, health care systems, and unmanned airborne vehicle (UAV) systems [2–4]. There has been extensive research and substantial advancements demonstrated in the various types of reconfigurable antennas made from rigid and conventional materials in the last few decades. New and modern communication systems' are moving toward wearable technology and system [5]. Wearable technology and systems include flexible electronics and displays, conformal consumer electronic gadgets and devices for body area networks (BANs) [6–8].

The requirements for flexible and reconfigurable antennas as a key component of the wearable systems have been increased because wearable devices must cope with the dynamics of the various surfaces.

Antenna reconfigurability is achieved by changing the antenna's current path, physical structure, or electrical properties. As Figure 1 shows, there are various methods to achieve reconfigurability in the four main categories. The progress in the development and realization of the flexible reconfigurable antennas was not as fast as their rigid counterparts. There are various existing challenges for the realization of these types of antennas. Figure 1 presents the existing reconfiguration techniques, which include: the PIN diode, the microelectromechanical systems (MEMS) switch, the varactor from the electrical category, the microfluidic-based method and the origami-based antenna. The photoconductive switch is made of semiconductor materials and uses bias lines instead of metallic wires [9]. The reason for not having a flexible reconfigurable structure with this switch is because of the challenges of integrating it with unconventional and flexible materials. Moreover, the limited range of flexible materials and permittivity restricts researchers in this area in employing reconfiguration techniques based on smart materials. These are based on reconfiguring the substrate of the antenna through changing its permittivity or permeability under different levels of voltage [10].

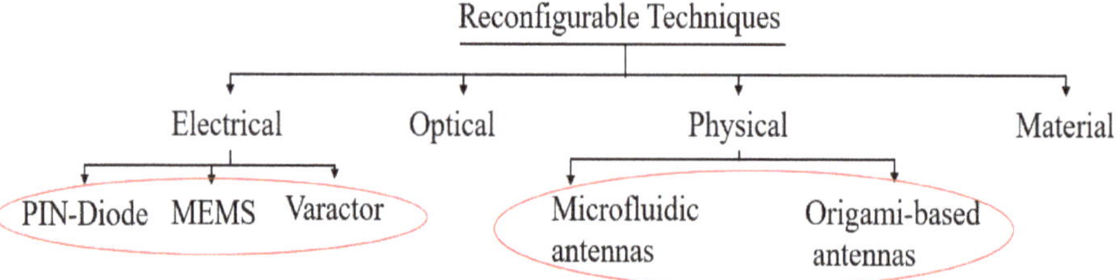

Figure 1. Categorization of reconfiguration techniques for all types of antennas.

A comprehensive review papers focusing exclusively on simulation [11], a specific reconfigurability method [10,12], antenna types [2,13], and other [14–16] of rigid antennas have been presented. Despite great efforts for comprehensive reviews of rigid antennas, there are no comprehensive review papers for the flexible reconfigurable antennas. In this review paper, we summarize recent developments in flexible pattern, frequency, and polarization reconfigurable antennas, as well as antennas with two feature reconfigurability. The purpose of this paper is to review the reconfiguration techniques, the challenges and limitations for design and fabrication, the achieved results and a comparison of the materials used in flexible and rigid structures. This leads to the recognizing of the current gap in this field and identifying of the possible future work in this area.

In the following sections, we discuss reconfigurable antennas in different categories:

- Frequency-reconfigurable antenna;
- Pattern-reconfigurable antenna;
- Polarization-reconfigurable antenna;
- Multiple reconfigurable features in an antenna.

2. Frequency-Reconfigurable Antennas

Frequency reconfigurable antennas have attracted researcher and industry communities' attention due to its capability to reduce the size of the front end systems and minimize interference with other wireless systems and maximize throughput [1]. The resonance of a proposed flexible antenna can be switched through the change of the effective length of radiator and the modification in the ground plane and feeding networks with the help of an electrical switch [17–25], by microfluidic-based antenna [26,27] and the use of a conductive fluid as a switch for a main radiating element [27], changing the antenna's permittivity [28,29] or the antenna's shape or type [30–53].

2.1. Antennas Based on Electrical Switches

Electrical switches are employed for the interconnecting of adjacent elements of the antenna patches [18–21] or varying the length of the slots on the patch or ground plane [1,22], modification on feeding networks [23] or switching between feeding ports [17] in order to realize frequency reconfigurability. Reconfigurability is limited to specific frequencies by employing a pin diode and MEMS switch, whereas continuous ranges of frequency reconfigurability are obtained by using varactor diodes.

In some research, the frequency reconfigurable antennas performance is analyzed through simulation [19,23] or it is verified through fabrication of prototypes that employ a 1 mm stub as an ideal diode. This is due to the challenges in the realization of the antenna with electrical switches such as placing diodes on unconventional materials, the need for a large RF bias network, and the effects of the nonlinearity of diodes and harmonics parameters. In other research, the conductive parts of the antenna where the diodes are placed are made of conventional materials to make the integration process easier [20,22,54].

In Reference [18], the microstrip patch antenna is placed on the Indigo blue jeans. A diode placed between two rectangles on the patch creates the reconfigurability between 2.4 GHz and 2.5 GHz for Bluetooth and WiMAX applications. In other research [23], two diodes are placed across a T-slot developed on the feed line of the rectangular patch antenna to reconfigure its frequency bands between 1.8 GHz, 2.3 GHz, and 2.4 GHz. In simulation analysis, Shieldlt Super is employed for the antenna conductive parts and it is placed on a Felt substrate with $\varepsilon_r = 1.44$. In Reference [19], dual-band reconfigurability is achieved through the connection of the two rectangular patches by eleven switches. An antenna is placed on the denim jeans textile and reconfiguring, between its two frequency bands 2.44 GHz (Wi-Fi) and 3.54 GHz (WiMAX).

Further investigation has been done by fabricating designed flexible reconfigurable antennas, employing DC circuits, and the measurement. The diodes are placed on the antenna, and in some cases, DC circuits are integrated as well. In Reference [20], the proposed textile antenna reconfigures between 2.45 GHz and 5 GHz. The copper tape is used for the patch and ground plane, and a denim material is used for the substrate. The diode is integrated on the copper tapes and connects two adjacent patches to create these two frequency bands. For the proposed antenna, the DC is connected to the antenna by means of two wires. In Reference [22], the proposed antenna is placed on a flexible polyethylene terephthalate which reconfigures between two states: a single band structure at 2.42 GHz and dual-band with different polarizations at 2.36 GHz and 3.64 GHz. The structure is composed of a folded slot and the radiation characteristics of the stub is altered by a PIN diode placed on the stub. The antenna is fabricated with inkjet-printing technology. Further study has been done on this antenna in [54]. An artificial magnetic conductor (AMC) surface is added to this structure in order to improve its radiation performance and to reduce the specific absorption rate (SAR). The performance of both antennas has been measured for the flat and curved positions. To control the diodes, DC voltage is applied with the RF signal through the input connector to eliminate the need of complex DC biasing circuit, thus preserving the antennas flexibility.

The same approach for biasing of the diode is employed in [24], but more challenges arise in the design process when unconventional materials are employed for the conductor and ground plane of the antenna.

The proposed antenna in [24] is fabricated by embedding conductive fabric into the flexible polydimethylsiloxane (PDMS) polymer [55–57] and all the active and passive elements of the antenna are fully encapsulated in the structure (Figure 2). To realize the antenna reconfigurability, two varactors are placed between a rectangular patch and a parasitic patch (made of conductive fabric) and are biased with same DC voltage. The frequency resonance is tuned continuously between 2.3 GHz and 2.65 GHz. The antenna in [21] consists of a rectangular-grounded loop which is excited by a monopole antenna fed by a coplanar waveguide. A U-shaped slot is loaded to the monopole to achieve dual-band operation.

This dual polarized antenna operates within 2.21 GHz to 2.69 GHz and 3.14 GHz to 3.55 GHz bands. The resonance of the higher band is tuned by switches placed within the U-slot.

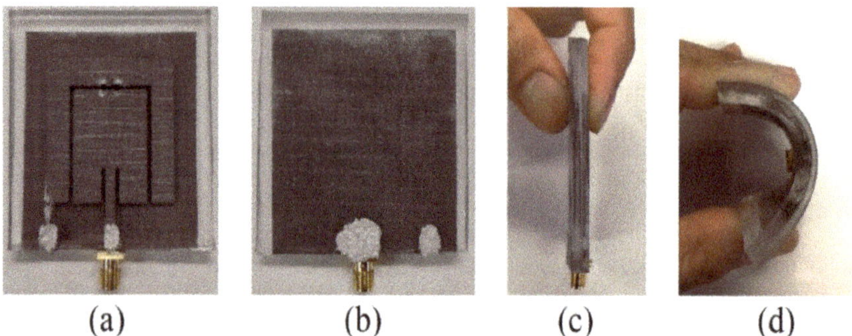

Figure 2. The fabricated frequency-reconfigurable antenna on polydimethylsiloxane (PDMS) substrate in [24]: (**a**) front view; (**b**) back view; (**c**) side view; (**d**) bending position.

Another approach for designing a reconfigurable antenna is by switching between two modes of the antenna through two-port excitation [17]. The proposed graphene conductive ink printed textile-based microstrip antenna in [17] can switch between 3.03 GHz and 5.17 GHz. The frequency reconfigurability is achieved by exciting the two modes of TM_{02} and TM_{20} through two-port excitation. A challenge of the switching between these two ports is addressed by introducing a single pole single throw switch to the structure.

DC circuits are integrated in different parts of antenna such as ground plane [1,58] or within the structure. In References [25,59], DC control circuits are integrated within commercial snap-on buttons. The snap-on reconfigurable wearable antennas are proposed to address the challenge of connection between rigid components to textile materials. This dual-band reconfigurable textile patch antenna is fabricated and the measurement results are reported showing a tuning range of 32.8% and 8.8% at 2.45 GHz and 5.8 GHz ISM bands, respectively [25].

The methods discussed so far have all been based on pin diodes and varactors. RF MEMS switches are also widely employed by researchers for realization of reconfigurability due to their low loss and distortion, small size, and high isolation [60]. The MEMS switch geometry and materials can be tailored to meet a specific size requirement and the desired actuation voltage, respectively [60,61]. While in most published literature, MEMS switches are integrated on the rigid substrate and one switch topology is used, great effort in integrating MEMS switches on flexible substrate is demonstrated in [62]. A multiband Sierpinski fractal antenna is proposed in [60], which is realized on a liquid crystal polymer (LCP) substrate, with three sets of used RF MEMS switches. Each switch has a different actuation voltage. The antenna operates at four frequencies between 2.4 GHz and 18 GHz without the need of extra lines to bias the switches.

As shown above, the main challenges for realizing frequency reconfigurable antennas with the electrical switches are based on the need for a DC-bias, and the adverse effects of rigid components on the antenna's flexibility. Moreover, in reconfigurable antennas based on electrical switches, antenna performance is degraded. The harmonics and intermodulation distortion is generated because of diode nonlinearities and gain compression.

2.2. Mechanically Frequency-Reconfigurable Antennas

Aside from the aforementioned challenges, the mechanical stability and rigidity of the frequency reconfigurable antennas based on electrical switches (PIN diodes and varactors) are not sufficient for some applications such as wearable antenna for gadget [63]. The high strength fields in MEMS switches limit the antennas with this type of switch to low-power applications [27]. To overcome these issues, the new categories of stretchable antennas such as microfluidic antennas, antennas based on micromesh structures, and origami-based antennas have been introduced. In the following sections, different categories of the mechanical frequency reconfigurable antennas with their existing challenges are presented.

2.2.1. Microfluidic Antennas

Recent development in the composite structure manufacturing creates the ability to embed microvascular networks in flexible materials and enables conductive or dielectric fluid to be injected into these materials. To reconfigure the antenna's frequency, the conductive fluid is used as a switch [27], the reactive loading effects of the fluid metal are used in [64,65], antenna fabricated by injecting alloys is elongated via stretching [66,67], the permittivity range of substrate is changed [28,29].

In microfluidic antennas in which conductive fluid is used as a switch, PDMS and styrene ethylene butylene styrene (SEBS) are usually used to produce microfluidic channels. The examples of liquid metals are mercury and eutectic gallium indium alloy (EGaIn). There is no need for electrical actuation switches in these types of the antenna which leads to enable efficient power handling. In the first microfluidic reconfigurable antennas with simple geometry, the liquid metal conductor is used in the PDMS channel [26]. In the presented frequency reconfigurable dipole antenna, for the microfluidic channels, SEBS with the $\varepsilon_r = 2.3$ and loss of 0.07 is employed and the channels are filled with EGaIn. The proposed antenna reconfigures from 2 GHz to 5.5 GHz by increasing strain up to 120% [68]. In Reference [27], the liquid metal is employed as a shortening switch. The frequency of the proposed antenna shifts between 4.6 GHz, 3.84 GHz and 5.34 GHz by emptying and filling the liquid metal to the microchannel to the lengths of $\lambda/2$ and 4λ. The antenna is placed on an S-glass composite substrate with $\varepsilon_r = 3.4$ at 4.6 GHz. A copper-tape and an EGaIn is used for the ground plane and the fluidic conductor, respectively.

In other research, the reactive loading effect of the fluid metal is used for tuning of a coplanar waveguide (CPW) folded slot antenna and miniaturization as well [64,65]. Two pairs of microfluidic channels are used as a switch to reconfigure between 2.4 GHz, 3.5 GHz, and 5.8 GHz [64]. The antenna is placed on Taconic TLY substrate with $\varepsilon_r = 2.2$, $\tan \delta = 0.0009$, with a thickness of 1.53 mm and two separate PDMS structures are used for two channels. The same approach is employed in [65], where liquid metal is also used as a reactive load on top of the antenna to achieve tuning and miniaturization as well.

In other types of microfluidic antennas, the frequency is tuned mechanically by elongating the antenna via stretching [66,67]. In Reference [66], a dipole antenna is fabricated by injecting EGaIn into microfluidic channels placed in PDMS substrate. The antenna frequency is tuned from 1850 MHz in a relaxed position (54 mm length) to 1600 MHz in the elongated position (66 mm length). In Reference [67], a stretchable frequency reconfigurable antenna is fabricated by injecting liquid metal alloy into a flexible substrate. Galinstan and silicone TC5005 are used for alloy and substrate, respectively. Instead of a microchannel, a square reservoir is fabricated in the flexible substrate. The frequency of the antenna is tuned from 1.3 GHz to 3 GHz by stretching up to 300% and varying the electrical length of the patch from 31 mm to 87 mm.

Despite the presented works [26,27,64–67] in which conductive fluid is used for radiating element, in [28,29], a microchannel is employed for adding different dielectric liquid materials to the antenna. The frequency shift is achieved in [28,29], by changing the permittivity range by inserting different fluids (air, acetone, and de-ionized (DI) water) into the existing channels.

As mentioned previously, fluidic-based antennas resist various or long time deformation compared to conventional antennas. In addition, the higher conductivity of conductive liquid compared to conductive fabrics, makes the design of an antenna with higher efficiency possible. On the other hand, there is a limitation for realizing the complex patterns for patch with this method. For antennas in which reconfigurability is achieved by changing the permittivity, the limited range of liquid dielectric, does not allow the researcher to have a wide range of reconfigurable antennas and frequency bands.

Moreover, their ability for stretching up is only in one direction for complex geometries, or in some cases the accurate mechanically control of antenna is difficult [69].

2.2.2. Antennas Based on Metallic Micromesh

In Reference [63], a semitransparent and flexible antenna which consists of a series of tortuous micromesh structures is presented. The proposed antenna can be stretched up to 40% without any problem. The mesh parts are made of copper and are placed on PDMS. The antenna's frequency reconfigures from 2.94 GHz to 2.46 GHz by increasing the tensile strain through mechanically elongating the meandering line of the antenna.

2.2.3. Kirigami and Origami-Based Antennas

In the class of mechanically reconfigurable structures, Kirigami and Origami-based antennas demonstrate RF characteristics that can be self-tuned by shape reconfiguration.

These antennas with their unique features are able to change the shapes and sizes of the 3-D structures or transform 2-D shapes into other 2-D and 3-D shapes by employing flexible materials [70,71]. This leads to the changing of the effective lengths [30], size [31] and type of antenna [32,33]. Various types of frequency reconfigurable origami-based antennas such as helix antennas [31,34,35], microstrip [32], yagi [36], dipole [37–39] and monopole [40] antennas have been presented in the literature.

In most of the existing 3-D origami helical antenna [41–49], accordion antenna [51], and spring structure [50], the frequency reconfigurability is achieved by changing the antenna's height. In [49], the proposed helical antenna is fabricated by copper foil on a paper, which switches between 1.82 and 2.14 GHz in folded state (H = 38 mm), and 860 MHz in the unfolded state (H = 255 mm). The same approach is presented in [43], in which origami quadrifilar helical antenna reconfigures between the normal mode at frequencies of 0.83 GHz, 1.17 GHz, or 1.5 GHz with H = 371 mm and axial mode at frequency of 1.23 GHz with circular polarization with H = 200 mm. In Reference [31], the frequency is tuned in the 3-D origami bifilar helical antenna by applying a force at its top and controlling its height. The antenna is made of copper tape on a sketching paper substrate with $\varepsilon_r = 3.2$. The frequency can be reconfigured among seven frequencies from 0.86 GHz to 3 GHz for three heights of 273 mm, 154 mm, and 25 mm. Another origami quadrifilar helical antenna is presented in [44]. The resonance frequency is tuned between 2.07 GHz, 3 GHz and 4.45 GHz by varying the height of antenna to 120 mm, 75 mm and 105 mm. Unlike the previously presented origami helical antenna, an origami reflector is used around this antenna to enhance its gain. This design is modified and an origami reconfigurable quadrifilar helical with circular polarization for higher frequency bands is presented [72]. The proposed antenna reconfigures between K, Ka, and extremely high-frequency bands by varying the height of antenna to 364 mm, 345 mm and 181 mm, respectively. Beside the helical origami structures, in spring antenna, the frequency can be tuned by changing the height of structure. In Reference [50], a tunable origami spring antenna with continuous frequency shifting from 1.1 GHz to 1.4 GHz is presented. The antenna is folded and unfolded with the help of an actuation system to the various heights from 60 mm to 145 mm. The actuation system is also employed in multi-radii monofilar helical antenna [73]. The antenna reconfigures between three bands of 1.22–1.84 GHz and 1.78–3.54 GHz and 3.64–4.04 GHz, with circular polarization. The antenna consists of a combination of two helical antennas with different radii and is fabricated from Kapton and copper trace.

Frequency reconfigurability based on converting 2D structure into 3D shape is presented in [30,32,37]. In References [30,37], reconfigurable origami-based dipole antennas with mechanically robust and durable structures have been introduced. In Reference [37], an E-textile accordion-dipole is fabricated on an Organza fabric. The frequency is reconfigured from 760 MHz to 1015 MHz range through changing the effective lengths by placing the antenna on a series of four Styrofoam fixtures which provide the changing angles of inclination from 0 to 60 degrees. A self-folding origami antenna that converts 2D structure into 3D shape is proposed in [32]. The antenna is a polystyrene subassembly that reconfigures between a microstrip patch antennas with a bandwidth of 1.5 GHz to 3.5 GHz to a monopole antenna with a bandwidth of 2.0 GHz to 2.3 GHz. The reconfigurability is realized by the concept of light activation and the bending of the polymer sheet assembly toward the light and away from the second port of the microstrip transmission line. The antenna is fabricated using RT/duroid 5870 with $\varepsilon_r = 2.33$.

The transformation from a planar structure to a 3-D structures for frequency reconfigurability is employed in [33,38,40,74]. A frequency switchable monopole antenna which is designed on an origami magic cube is proposed in [40]. The antenna reconfigures between 850 MHz to 1000 MHz and 1.3 GHz to 1.6 GHz for the unfolded and folded states due to the change in the length of the antenna. The same approach is used in [74] to design a dual-band antenna operating at 1.57 GHz and 2.4 GHz in the folded state and 900 MHz and 2.3 GHz in the unfolded state. The proposed antenna consists of a meandered monopole that is loaded by a microstrip open-ended stub. The open-ended stub controls the second resonance of antenna by changing its length (Figure 3). In Reference [38], a frequency-reconfigurable dipole antenna using an origami flasher is proposed. In this structure, the dipole length is increased in the folded state into a cube. The antenna frequency is switched between 750–800 MHz (folded) and 1.19–1.26 GHz (unfolded). The transformation from a dipole structure to a 3D conical spiral structure is made in a proposed morphing Nojima origami antenna in [33]. The antenna operates at 0.48 GHz and 2.5 GHz in unfolded (dipole antenna) and fully folded state (conical spiral antenna), respectively.

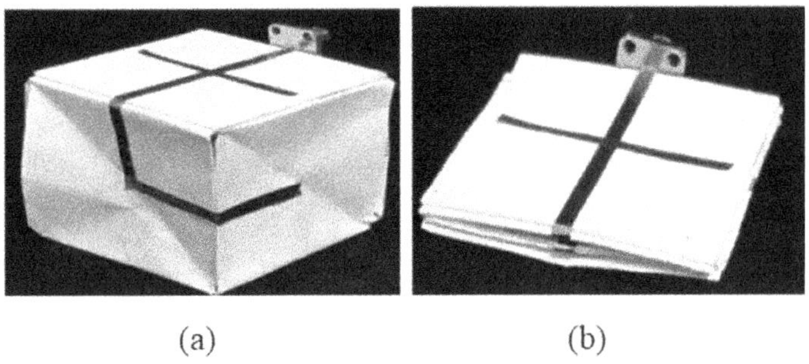

(a)   (b)

Figure 3. The frequency reconfigurable antenna based on changing the antenna shape (Origami Magic Cube): (**a**) unfolded; (**b**) folded [74].

Most origami-based antennas are designed using paper which creates poor durability and performance stability [75]. Moreover, to accurately control the change in the antenna dimensions and shape is not easy due to the high flexibility nature of paper. This also limits the range of the frequency reconfiguration to a couple of points. Therefore, in [75], a dipole antenna based on multi-material 3D printing technology is proposed, which can provide a continuous frequency reconfigurability from 0.95 GHz to 1.6 GHz through a precise structure manipulation by a robot.

In Reference [76] a frequency reconfigurable monopole antenna based on kirigami technique is presented. A three-story tower kirigami is used for this antenna. This technique provides a more stable structure compared to a free-standing structure, as well as a volume reduction. In this antenna, a horizontal monopole antenna is designed on a two dimensional flat PET film with a frequency of 0.42 GHz, which can be stretched to the three-storey tower to operate at a frequency of 0.53 GHz.

Most of the origami-based antennas use paper in their structure due to its high foldability, high thermal stability and lightweight. Despite the advantages of using paper in origami antennas, the roughness on the surface of paper creates some electrical properties of the conductive patterns. This can be addressed by coating the paper with polymers, but the folding of antenna will be affected. Therefore, in [77], a nonopaper approach with higher transparency and a smoother surface is presented for a foldable frequency reconfigurable antenna. The frequency of V-shaped silver nanowire antennas on nonopaper is switched between 2.3 GHz to 3.0 GHz in unfolded and folded states, respectively.

One of the challenges of this type of antenna is its limited durability. The conductive parts of these antennas such as copper tape and conductive inks are prone to delamination and deterioration in conductivity due to the extensive mechanical stress. In addition, substrates such as paper and memory polymers are not robust to extensive folding [37]. Although origami-based antennas are not practical for communication systems yet, they have the potential to be used with advances in material and actuator technologies [38].

In Table 1, the presented flexible frequency-reconfigurable antennas are listed. The table presents the antenna types, reconfigurability method, number of states, radiation performance and the materials that were used in these antennas to give a general overview on this category.

2.3. Antennas at Millimetre-Wave

The millimeter-wave (MMW) spectrum addresses the challenges of the demands for high bandwidth to cover high data rates and extends capacity to facilitate the services and applications for 5G. An adaptable and flexible antenna at wireless front-ends is required for the realization of 5G mm wave architectures [78].

As mentioned earlier, one of the challenges in designing flexible reconfigurable antennas is the choice of the substrate. This becomes more challenging in mmW because of the higher dielectric loss [55]. LCPs and polyethylene terephthalate (PET) which feature low loss-tangent, good flexibility, conformity, and surface adhesion are presented as great candidates for the substrates at higher frequency bands [79]. In addition, the inkjet-printing as a cost-effective, and precise fabrication method is a promising candidate due to its sensitivity of dimensions (very low wavelength) in mmW and its capability of fabricating of thin conducting layers. In Reference [78], a slotted T-shaped radiating patch embedded in a rectangular aperture cut inside the ground plane is inkjet-printed on a PET for operating frequency bands between 26.5–40 GHz. Four ideal switches, open/short metal contacts are employed in the middle of slots to generate the four modes of reconfigurability in this antenna.

In Reference [79], a flexible frequency reconfigurable antenna in 20.7–36 GHz with four different switches configurations for 5G networks is presented. The antenna is placed on LCP and consists of a radiating fork shaped-patch and two stubs which are connected to the main patch by means of two PIN diodes. The inkjet printing for the conductive parts of the antenna which include the patch, stubs, and matched coplanar waveguide (CPW) feeding structure are employed. The efficiency of the antenna is reported to be in excess of 65% across the entire bandwidth.

Table 1. Comparison of methods and materials used in different kinds of flexible frequency-reconfigurable antennas, along with corresponding performance.

Ref.	Antenna Type	Reconfigurability Method	No of States	Used Materials	Freq. (GHz)		Gain (dBi)		Eff. (%)		BW (%)		Dimension ($\lambda_{min} \times \lambda_{min}$)	Height (λ_{min})
					Min.	Max.	Min.	Max.	Min.	Max.	Min.	Max.		
[1]	patch	electrical switch/slot modification on ground	6	ShieldIt Super and Felt	1.57	2.55	0.2	4.8	17	47	2.9	19	0.59×0.51	0.006
[20]	patch	electrical switch/connecting two patches	2	copper tape, Denim	2.45	5	3.17	3.55	N/A	N/A	2.5	5.85	0.40×0.19	0.008
[21]	loop/monopole	electrical switch/geometry morphing	3	transparent PETP	2.4	3.4	1.9	3.2	82.8	97	12.25	20.94	0.264×0.4	0.001
[27]	dipole	physical/microfluidic based	2	S-glass, copper-tape, EGaIn	3.84	5.34	N/A	N/A	N/A	N/A	N/A	N/A	1.8×1.8	N/A
[24]	patch	electrical switch/parasitic patch	Continuous	Polymer, Conductive fabric	2.3	2.68	2.9	3.3	40.3	46.1	0.3	0.4	0.4×0.43	0.042
[28]	slot antenna	physical/change of permittivity by injection	3	FR4, PDMS, copper	3.05	7.9	N/A	N/A	N/A	N/A	N/A	N/A	N/A	0.015
[31]	helical	physical/height of structure	3	copper tape, sketching-paper	0.86	3	4.69	6.79	N/A	N/A	N/A	N/A	0.13×0.13	0.07, 0.44, 0.78
[32]	monopole/microstrip	physical/changing the type of antenna	2	copper-clad RT/duroid 5870	1.5	3.5	N/A	N/A	N/A	N/A	13.9	80	N/A	N/A
[50]	Spring Antenna	physical/height of structure	6	Copper tape, paper	1.1	1.4	9	~10	N/A	N/A	N/A	N/A	N/A	0.22 to 0.53
[66]	patch	physical/based on injection alloys, reconfigurability by stretching up	Continuous	EGaIn, PDMS	1.6	1.85	N/A	N/A	N/A	N/A	12.5	16.5	0.053×0.29 to 0.053×0.35	0.005

3. Pattern-Reconfigurable Antennas

Pattern reconfiguration is an interesting characteristic of antennas and is often used of changing the coverage from one area of interest to another one, switched operation between two or more wireless nodes, determining direction of arrival, increasing or decreasing the coverage area of a wireless node, or more generally, to adapt the performance of a wireless node in response to varying conditions in surrounding environment. Thus, our discussion of pattern reconfiguration includes antennas that can reconfigure the radiation pattern to two (a special case often also referred to as "switchable antennas") or more states. In the presented flexible pattern reconfigurable antennas, the reconfigurability was achieved by using an electrical switch by changing the ground or radiator shapes [80–84], switching between two types of the antenna combined in a main structure [85], shorting the radiator to the ground plane [86,87], switching between feeding ports [88] or physical reconfiguration by changing the antenna's shape or type [89–95].

3.1. Pattern-Reconfigurable Antennas Based-On Electircal Switches

For pattern reconfigurable antennas based-on electrical switches, the reconfigurability is achieved by connecting the adjacent radiators or parasitic elements to the main patch, switching between two types of the antennas combined in the main structure. In Reference [81], the four-element circles are located at 0°/+360°, +90°, +180° and +270° and are connected by four arms to the center. The antenna's patterns are reconfigured to four different directions of 0°, +90°, +180°, and +270° by using four RF switches which are placed on the arms between the center and the radiation circles. The antenna's conductive parts are made of ShieldIt Super textile and are placed on Felt as a substrate. A similar approach has been used with the same research group in [96]. The antenna consists of a centre patch and four radiating elements located at 0°/+360°, +90°, +180° and +270° which are connected to the center by four PIN diodes. The same material in [81] is employed in this antenna as well. The radiation pattern of antenna is doughnut-shaped and the maximum gain is at 0°/+360°, +90°, +180° and +270° based on switches' statuses. The antenna in [82] comprises of a circular monopole patch, three parasitic patches, and two parallel symmetrical parasitic branches embedded with two PIN diodes which operates between 2.40 GHz and 2.48 GHz. The antenna's pattern is reconfigured to the beam in the direction 30°, 330° in the azimuthal planes by controlling the two diodes. The antenna is fabricated on Rogers 5880 substrate with $\varepsilon_r = 2.2$ and a thickness of 0.127 mm.

In Reference [85], a bent dipole and a loop are combined in one structure and the radiation patterns of the two antennas are canceled or compensated by using two diodes. This antenna was fabricated on flexible polyimide with $\varepsilon_r = 3.6$ and loss of 0.01. The beam direction is steered to +50°, 0°, and −50° in the azimuth direction at 2.5 GHz (Figure 4). In Reference [97], a top-loaded monopole antenna and a loop antenna are combined in a one structure. Three antennas with different top-loading angles are proposed and two diodes are placed on antenna. For antenna with angle of 90°, the beam is steered to two directions of −90° and +90°. The antenna is fabricated on FR-4 substrate and operates at 2.4 GHz for Fitbit Flex Wristband.

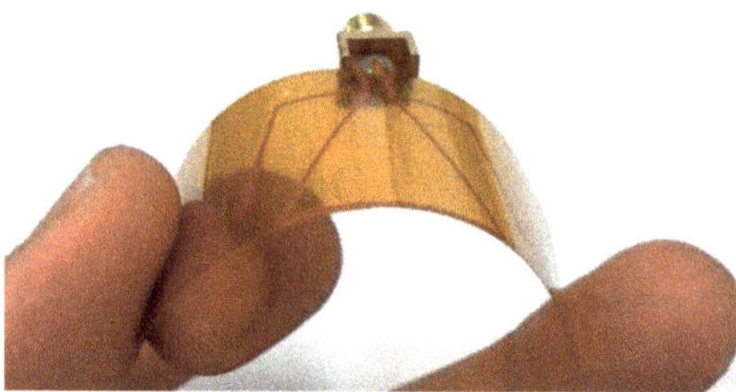

Figure 4. Fabricated prototype of the pattern-reconfigurable antenna on flexible polyimide in [85].

3.2. Mechanically Pattern-Reconfigurable Antennas

Mechanically pattern reconfigurable antennas are presented in response to disadvantages of using electrical switches, which were discussed in previous section. A pattern reconfigurable origami quasi-Yagi helical antenna using a copper film on DNA shaped PET substrate is presented in [91]. The pattern is reconfigured between four states at 2.2 GHz by changing the role of the DNA reflector and the DNA based director. In this antenna, three origami DNA geometries are used. Besides the reflector and the director, the antenna has a DNA shaped driven element. In States 1 and 2 of the antenna, the beam is directed to 30° and −30° where a parasitic director is designed by folded origami DNA and parasitic reflector is made of unfolded origami DNA for State 1 and vice versa for State 2. In State 3, the beam is directed to 0° and in State 4 it is directed at −50° and 50°, where both the parasitic elements are folded in State 3 and unfolded in State 4.

A thermally reconfigurable antenna based on origami reflectors with ability to switch the beam direction and beamwidth is presented in [92]. The proposed antenna comprises of a single monopole antenna and four origami reflectors made of copper-clad FR4 substrate. Smart shape memory polymer hinges are used to stimulate the reflectors. The beam of antenna is switched between omnidirectional pattern in State 1 (all reflectors are folded), directional pattern in the 45°, 135°, 225°, and 315° in State 2 (by unfolding various reflector pairs) and directional beam in 0°, 90°, 180°, and 270° in State 3 (by unfolding each origami reflector). In addition, the 3 dB beamwidth is changed from 52° (State 2) to 140° (State 3).

The antenna's half-power beamwidth (HPBW) is changed from a 360° to 40° in a bi-directional pattern with a loop antenna array based on magic cube origami in [95]. The antenna is an array which consists of three single cubes with a single loop antenna, in serious form. The array can be folded and unfolded due to the nature of structure. The single antenna on a single cube was designed at 1.39 GHz with 4.03 dBi gain and 360 degrees HPBW. The series of two and three loops by using two and three cubes is increased to 5.2 dBi (HPBW = 60) and 5.53 dBi (HPBW = 40), respectively. Figure 5 shows the fabricated antenna for three states. In another quasi-Yagi monopole antenna based on origami magic spiral cubes, the pattern is reconfigured between omnidirectional radiations and three other broadside patterns at 1.9 GHz [93]. The antenna consists of a stacked L-shaped driven monopole, an L-shaped reflector on the first origami magic cube, and two L-shaped directors on the Cube 2 and 3 3, all are fabricated on paper. In the first state, the monopole antenna has omnidirectional pattern with 1.9 dBi peak gain, which is increased to 5.7 dBi by adding L-shaped reflector. By adding first and second directors to this antenna the gain is increased to 6.8 dBi and 7.3 dBi, respectively.

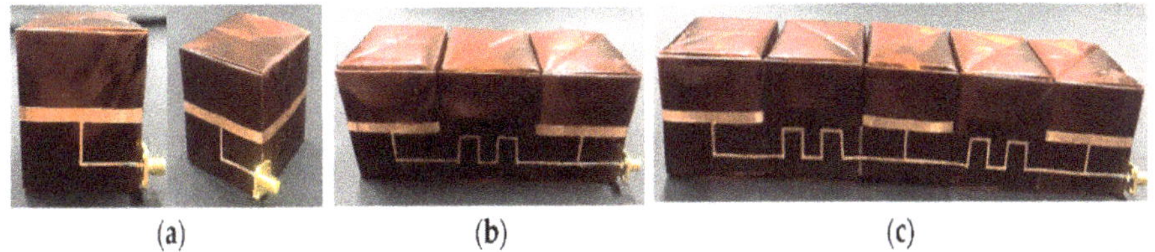

Figure 5. Fabricated proposed antenna array based on magic cube origami for: (**a**) single antenna and cube (**b**) two antennas and (**c**) three antennas [95].

One of the challenges in origami-based pattern reconfigurable antennas is to maintain the shape after being folded or unfolded. In Reference [98], a suitable Kresling conical is presented for a spiral antenna. This antenna shape can be kept by its internal tension after unfolding. The pattern of this antenna is reconfigured between omnidirectional (folded) for planar spiral and unidirectional for conical spiral shape (unfolded) with circular polarization at 2.5 GHz to 3.2 GHz. This challenge is also addressed in [94]. A pattern-reconfigurable axial-mode helix antenna by controlling the height of helix is presented. The height is controlled by applying a direct current to the SMA spring, which is an actuator.

In Table 2, the flexible pattern-reconfigurable antennas are listed. The table presents the antenna types, reconfigurability method, number of states, the radiation performance and the materials that were used in these antennas.

3.3. Dual Mode Antennas

Recent advances in communication technology have led to the advent of antennas for body area networks (BANs) applications. The suitable radiation pattern for connecting nodes that are placed on-body is monopole-like patterns and connecting nodes that are placed on-body and off-body are broadside patterns [88]. New types of the flexible dual-mode antennas which reconfigure between on-body and off-body radiation patterns have been presented [88–90,99,100].

In Reference [88], a flexible reconfigurable dual-mode microstrip antenna operating at 2.45 GHz is presented. A ring and a meandered patch with two different feeding ports and the same ground plane are combined in one structure for the on-body and off-body modes, respectively. The pattern of the antenna is changed by switching between two feeding ports, the monopole-like of the ring pattern and a broadside pattern of the meandered patch. The pure copper polyester taffeta fabric is used for the conductive parts of the antenna and it is placed on felt with $\varepsilon_r = 1.17$ and loss of 0.016. The effects of the body tissue on antenna performance is investigated by placing an antenna on a phantom.

The dual-mode feature by having broadside radiation pattern at 2.5 GHz and a conical-like radiation pattern at 5.8 GHz in the triangular patch antenna is achieved [99]. A conductive textile is used for the patch and the ground plane and antenna is placed on the substrate with $\varepsilon_r = 1.63$ and loss tangent 0.04. The reconfigurability is achieved by using three vias and two symmetrical open-ended slots. The proposed flexible microstrip patch antenna in [100], reconfigures between on and off-body communications at 2.45 GHz by employing two PIN diodes. The textile material felt and a conducting fabric pure copper taffeta is utilized for dielectric layers and metallic layers, respectively.

As it is mentioned before, mechanical tuning techniques including physical deformation, stretching, or folding for pattern reconfigurable are presented as an alternative to electronic tuning [90]. A mechanical pattern reconfigurable antenna based on liquid metal and microchannel inside a polymer is presented in [90]. This antenna comprises a 9×5 array of metal dipoles which are inside platinum-cured thermoset silicone with $\varepsilon_r = 2.2$ and loss of 0.016. To form the dipoles, EGaIn is injected into the channels inside the substrate. Pattern reconfigurability is achieved through elongation from 0 to 53% to theta 0 to ±55° and surface reshaping to 0, ±37 and ±58 for flat, convex and concave, respectively. In a fully textile reconfigurable ultra-wideband (UWB) antenna, the pattern is switched between a monopole-like shape from 2.62 GHz to 10.1 GHz and a microstrip-like shape from 2.73 GHz to 10.1 GHz [89]. The reconfigurability in this antenna is achieved by raising and lowering a textile flap which is the main radiator. As Figure 6 shows, when this textile flap is parallel to the ground plane, the structure behaves like a microstrip antenna with a unidirectional pattern. When in the orthogonal position to the ground plane, the structure acts as a monopole antenna with an omnidirectional pattern.

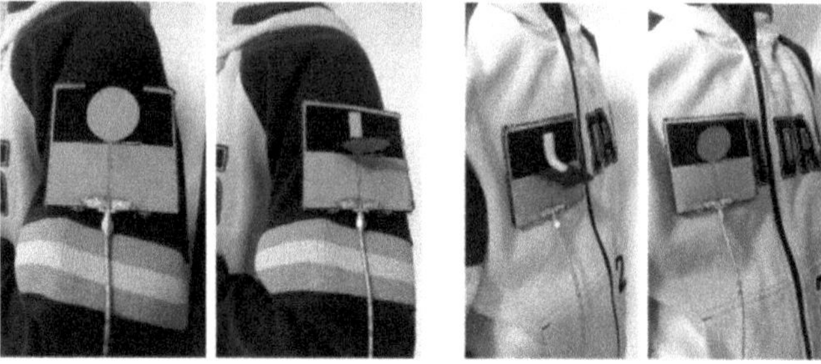

Figure 6. The measurement set-up for proposed fully textile reconfigurable UWB antenna with two monopole-like and a microstrip-like patterns ([89], reproduced courtesy of The Electromagnetics Academy).

Table 2. Comparison of methods and materials used in different kinds of flexible pattern-reconfigurable antennas, along with corresponding performance.

Ref.	Antenna Type	Reconfigurability Method	No of States	Freq. (GHz)	Used Materials	Gain (dBi)		Eff. (%)		BW (%)		Dimension ($\lambda_{min} \times \lambda_{min}$)	Height (λ_{min})
						Min.	Max.	Min.	Max.	Min.	Max.		
[81]	patch	electrical switch/connecting parasitic patches to the main patch	4	2.45	ShieldIt Super, Felt	N/A	N/A	N/A	N/A	N/A	N/A	0.72×0.72	0.016
[82]	monopole	electrical switch/connecting surrounding elements to the main patch	2	2.40–2.48	Rogers5880	0.5	0.7	77	78	N/A	3.2	0.28×0.2	0.001
[85]	dipole, loop	electrical switch/combination of the pattern of two antennas	3	2.47–2.53	Polyimide	1.96	2.48	N/A	N/A	N/A	2.4	0.2×0.35	0.004
[86]	patch	electrical switch/connecting patch to the ground	2	2.4	Felt, conductive textile	2	3.9	38	49.5	4.8	8.5	0.78×0.78	0.023
[88]	patch	electrical switch/switching between feeding ports	2	2.45	felt and Copper Polyester Taffeta	2.9	5.2	65	72	4.1	4.9	0.64×0.64	0.03
[90]	array of dipoles	Physical/elongation of patch which is based on liquid metal	3	12.5	EGaIn, platinum-cured thermoset silicone	N/A	N/A	N/A	N/A	N/A	N/A	N/A	N/A
[91]	helical	Physical/folding and unfolding	4	2.2	copper film, PET	7.5	9.5	N/A	N/A	N/A	N/A	N/A	N/A
[92]	monopole	Physical/folding and unfolding origami reflectors	8	2.4	copper-clad FR4 substrate	4.5	11	86	98	16	30	2.2×2.2 to 0.73×0.73	~0 to 0.73
[93]	monopole	Physical/folding and unfolding two other directors on magic cube	4	1.9	copper film, paper	1.9	7.3	64	N/A	N/A	21	0.24×0.24	0.06 to 0.72
[95]	loop	Physical/magic cube origami	3	1.39	Copper tape, Kapton	4.03	5.53	89	97	10	18	0.21×0.21 to 0.21×0.63	0.21

4. Polarization-Reconfigurable Antennas

The polarization mismatch between antennas in receivers and transmitters in communication systems creates power loss. The mismatch is because of the alteration of position and orientation of the two antennas. The polarization reconfigurability is a feature that compensates this power loss by matching the antenna's polarization [101]. Polarization reconfigurability can be achieved by changing the shape of the radiating patch [101], switching between different radiators [102] and changing the antennas shape [103–107].

In Reference [101], three polarization modes are created by controlling the truncations on the patch by using four diodes at 2.4 GHz. The proposed textile-based antenna is placed on felt substrate. The performance of the antenna over a phantom is analyzed to verify its suitability for BAN applications. In another proposed textile-based substrate-integrated waveguide (SIW) circular ring-slot antenna [108], the inclination of the linear polarization is controlled in the range of angles 30° by changing the position of the short strip to three places on the slot ring.

In Reference [102], by switching between four sector radiating elements located at 0°, 360°, 90°, 180°, and 270°, which are interconnected by pin diode switches to the center, the antenna exhibits dual-polarization with a combination of vertical and horizontal polarizations with a omnidirectional radiation pattern. The proposed antenna is placed on felt and conductive textile (E-textile) and is used for the ground plane and radiating elements. The polarization of the antenna is reconfigured between two circular states at 1.85 GHz to 2.75 GHz (unfolded, H = 106 mm) and a linear state at 1.4 GHz and 3.7 GHz (folded state, H = 53 mm) in an origami conical spiral antenna [107]. Another mechanical polarization reconfigurable antenna based on origami and skeleton scaffoldings is presented in [103–105]. The proposed bifilar segmented helical antennas reconfigure between right-hand circular polarization and left-hand circular polarization by retaining around their central axis (Figure 7) [103]. Paper with 100 μm thickness is used as the origami base for this structure. In other mechanical polarization reconfigurable textile patch antennas, the conductive Velcro tape and zip fastener is used separately to switch between linear to circular polarizations at 2.3–2.5 GHz [106]. Both antennas are made of felt and conductive cloth. Two antenna's axial ratio repeatability ability to interchanging polarization modes are compared in this research. The primary advantage of this research is its ability to adjust the antenna characteristics.

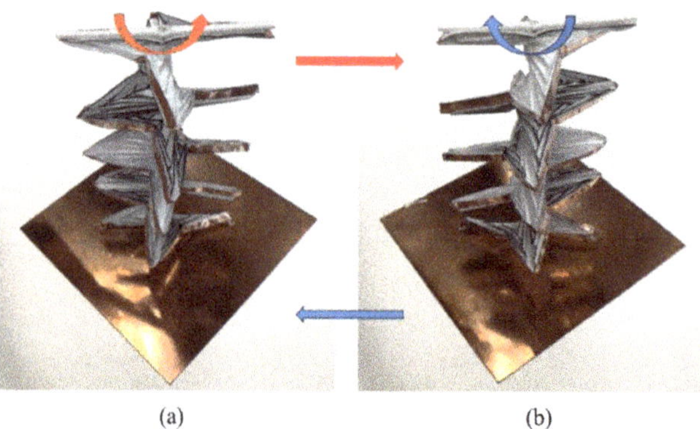

(a)   (b)

Figure 7. The prototype of the bifilar segmented helical antennas at (**a**) left-handed and (**b**) right-handed angles [103].

5. Multiple Reconfigurable Features in Antennas

In multi-reconfigurable antennas, the two features of the antenna can be switched independently. These antennas are listed in a couple of categories:

- Frequency and polarization;
- Frequency and pattern.

The advantages of reconfigurability of these two features compared to the one feature of the antenna is having more flexibility and diversity which leads to a new domain of research.

5.1. Frequency and Polarization Reconfigurability

In Reference [109], an L-shaped stub is introduced in the ground plane of a disc monopole antenna and a parasitic arc is placed around the main radiator. Two-pin diodes are placed between the main radiator and the parasitic arc to switch between two frequencies of 2.45 GHz and 0.92 GHz. The L-shaped stub is also connected to the ground plane with the help of two diodes. Through optimization of the diodes positions on the patch and the connecting and disconnecting the stub to the ground, four states of right-handed circular and left-handed circular polarizations are generated for both frequency bands, separately. The antenna is placed on jeans substrate with $\varepsilon_r = 1.7$ and an adhesive copper tape is used as a radiating element and ground plane (Figure 8). In References [1] and [58], the frequency reconfigurability is achieved by manipulating the slot size at the ground plane of the truncated edges of the rectangular patch textile antenna. In Reference [1], three diodes are integrated on a slot at the ground plane of a textile antenna to switch between six frequencies: 1.57 GHz, 1.67 GHz, 1.68 GHz, 2.43 GHz, 2.50 GHz, and 2.55 GHz. ShieldIt Super and felt with $\varepsilon_r. = 1.22$ at 1.575 GHz is utilized for conductive parts and substrate, respectively. To bias these three diodes, four vertical slots at the edge of the ground plane are added to the structure. Capacitors are employed on vertical slots to preserve the RF current flow on the ground plane. Four wires are connected to the antenna to provide voltage and to switch the diodes (Figure 9).

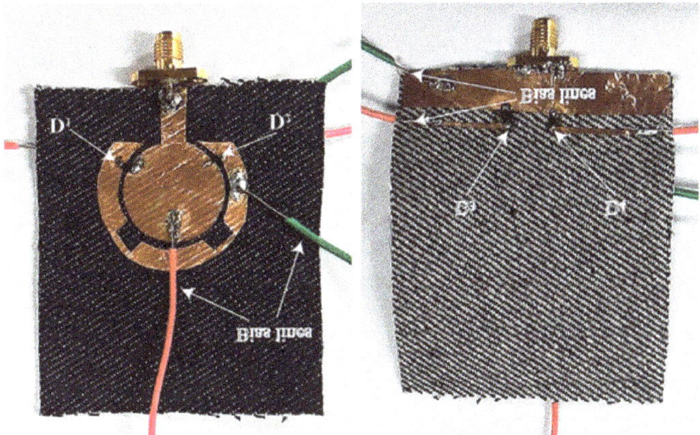

Figure 8. Fabricated prototype of the frequency and polarization-reconfigurable antenna on jean substrate in [109].

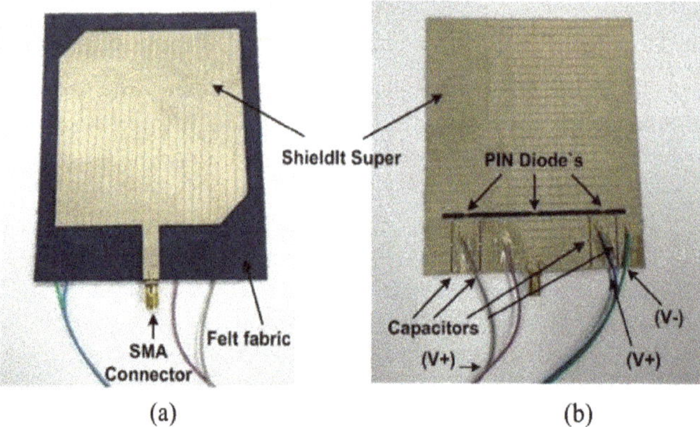

Figure 9. The fabricated frequency-reconfigurable antenna based on textile in [1]: (**a**) front view; (**b**) back view.

Polarization reconfigurability between right-handed and left-handed circular polarization at 5 GHz, and linear polarization at 4.7 GHz or 5.2 GHz is presented with a textile antenna based on snap-on buttons in [110]. The polarization is controlled by rotating the module through opening and closing the flap (the slot placed on the circular patch antenna). The metalized nylon Ripstop fabric and Cuming Microwave C-Foam PF-4 foam with $\varepsilon_r = 1.06.$ are used for conductive parts and substrate, respectively.

5.2. Frequency and Pattern Reconfigurability

The proposed antenna in [111] consists of two symmetrical radiating elements, with a shared feedline and a ground plane which is fabricated on a Rogers5880 substrate. Each radiation element comprises a circular monopole branch, a regular hexagonal open ring, and two other branches added to the ring. The frequency of the antenna reconfigures between two bandwidths 1.84–2.00 GHz and 2.27–2.49 GHz. It is able to steer the beam in two directions (60° and 300° for the lower band and 34° and 326° for the higher band) by employing 8 PIN diodes loaded on the symmetric hexagonal split ring and monopole branches. An ultra-wideband flexible frequency and pattern reconfigurable antenna that uses two fabrication methods of conductive thread embroidery on the textile substrate and cutting the thin sheet of copper with and sticking it on the substrate is presented in [112]. The frequency and pattern of the antenna are reconfigured for six frequency bands in a range between 2 GHz to 10 GHz and three radiation patterns modes by using 10 stubs on the front part and two semi-elliptical stubs located on the back part, respectively. In Reference [113], the antenna with both frequency and pattern reconfigurability is proposed. To achieve the pattern reconfigurability in this structure which can be changed between a straight line and helix with 1.1-turn, liquid crystalline elastomers and heating effects are employed on their shape. This material phase transition to an isotropic state happens through heating. By using the behavior of liquid crystalline elastomers and adhering thin metal, a reconfigurable antenna is designed. At 30 °C, the antenna is straight (Figure 10a) and has an omnidirectional pattern at 4.9 GHz. The frequency and pattern are shifted to 14.04 GHz and directional by increasing the temperature to 92 °C and having 1.1-turn helix (Figure 10b). This antenna has a capability of reconfiguring the frequency from 12 GHz to 10.7 GHz as well by changing the shape of antenna from a 0.5-turn loop to a 1.5-turn helix.

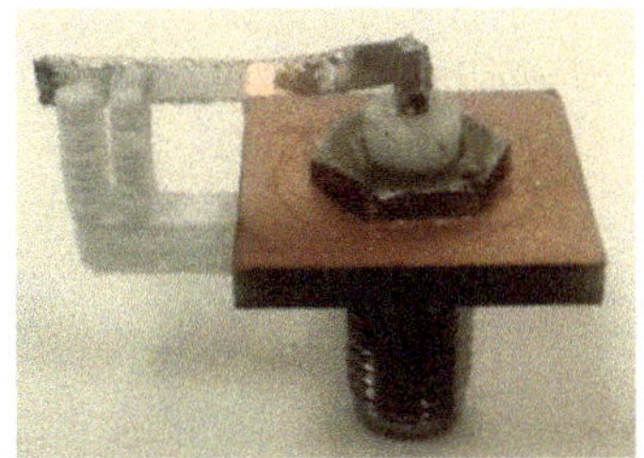

Figure 10. Pattern and frequency-reconfigurable antenna based on liquid crystalline elastomers: (**a**) straight statues; (**b**) helix structure with 1.1-turn [113].

Most of the origami-based antennas reconfigure frequency due to the nature of mechanical tuning which is based on changing antenna's dimension. The Nojima wrapping for origami-based structure was proposed for the first time in [114]. It provides the various models based on the shape of central hub, which is a square shaped in this paper, and different angles between the segments as well. In Reference [115], a dual-mode origami Nojima antenna which reconfigures between directional mode at 1.61 GHz for folded state and omnidirectional mode at 0.66 GHz for unfolded state is presented.

6. Comparison between Proposed Techniques for Flexible Reconfigurable Antennas

Table 3 shows a list of materials that are mostly used for flexible and conformal reconfigurable antennas is shown. The table provides the available materials for each fabrication method and their suitability for bending/folding and sustained bending/folding (repeating deformation) for better comparison. For all of frequency, pattern, polarization, or a combination of these characteristics, the reconfigurability methods can be divided into three categories: electrical switch, antenna based on liquid metal/dielectric and Origami-based antennas.

Among the presented techniques, antennas based on electric switches employ a much wider range of materials compared to microfluidic and origami-based antennas. Their frequency can be tuned to a couple of states or even a continuous range, whereas limitations exist for the other techniques. However, if actual diodes are employed instead of ideal stubs, a DC-bias and a controlling system are required. Therefore, the flexibility of the antenna will be affected by the integration of the antenna with rigid components. The strong connection between the rigid components and unconventional materials electrically and mechanically is also challenging. For this technique, the nonlinearity effects of diodes on antenna performance should be considered. Furthermore, the efficiency of these types of antennas is usually lower compared to their rigid counterparts. As Table 3 shows, this is due to the lower conductivity of the flexible materials compared to the conventional materials e.g., copper layer with conductivity of 5.96×10^7 S/m.

Table 3. Characteristics of materials with relevance to reconfiguration element used for flexible and conformal reconfigurable antennas along with corresponding performance.

Reconfiguration Element	Ref.	Material Name	Material Type	Dielectric Constant (Relative Permittivity) ε_r	Dielectric Constant (Relative Permittivity) $tan\,\delta$	Conductivity (S/m) or Resistivity (ohms/sq)	Suitable for Folding/Bending	Suitable for Sustained Folding/Bending	Freq. (GHz) Min.	Freq. (GHz) Max.	Gain (dBi) Min.	Gain (dBi) Max.	Eff. (%) Min.	Eff. (%) Max.	BW (%) Min.	BW (%) Max.	Dimension ($\lambda_{min} \times \lambda_{min}$)	Height (λ_{min})
	[1]	ShieldIt Super	Textile	-	-	1.18×10^5 S/m	Yes	No	1.57	2.55	0.2	4.8	17	47	2.9	15.4	0.59×0.51	0.006
		Felt	Textile	1.22	0.016	-												
	[17]	Graphene	Conductive Ink	-	-	0.37×10^5 S/m	Yes	No	3.03	6.13	0	2.09	54.9	74	N/A	N/A	0.46×0.54	0.015
		Polyvinyl chloride foam, polyimide film	Polymer, Foam	1.5,3.5	-	-												
	[20]	Copper tape	Copper	-	-	-	Yes	No	2.45	5	3.17	3.55	N/A	N/A	2.5	5.85	0.40×0.19	0.008
		Denim	Fabric	1.68	0.03	-												
	[24]	a nylon ripstop fabric	Fabric	-	-	0.01 Ω/sq	Yes	No	2.3	2.68	2.9	3.3	40.3	46.1	0.3	0.4	0.4×0.43	0.042
		PDMS	Polymer	2.82	0.023	-												
	[25]	Shieldex	Fabric	-	-	0.009 Ω/sq	Yes	No	2.1	6.2	−0.9	8.8	30	90	N/A	N/A	0.42×0.42	0.04
		C-Foam PF-4	Foam	1.06	0.0001	-												
Conductive fluid as a switch	[27]	EGaIn	Liquid Metal	-	-	3.4×10^6 S/m	Yes	Yes	3.84	5.34	N/A	N/A	N/A	N/A	N/A	N/A	1.8×1.8	N/A
		S-glass	Fibers & Textile	3.4	-	-												
Microfluidic antenna-physical deformation	[66]	EGaIn	Liquid Metal	-	-	3.4×10^4 S/m	Yes	Yes	1.6	1.85	N/A	N/A	N/A	N/A	12.5	16.5	0.053×0.29 to 0.053×0.35	0.005
		PDMS	Polymer	2.67	0.0375	-												
	[67]	Galinstan	Liquid Metal	-	-	3.46×10^6 S/m	Yes	Yes	1.3	3	N/A	N/A	65	80	N/A	N/A	0.17×0.65	0.02
		TC5005	Silicone	2.8–3.1	-	-												
	[90]	EGaIn	Liquid Metal	-	-	3.4×10^4 S/m	Yes	Yes	N/A	N/A	N/A	N/A	N/A	N/A	N/A	N/A	N/A	N/A
		Platinum-cured thermoset silicone	Polymer	2.2	0.08	-												

Table 3. *Cont.*

Reconfiguration Element	Ref.	Material Name	Material Type	Dielectric Constant (Relative Permittivity)		Conductivity (S/m) or Resistivity (ohms/sq)	Suitable for Folding/Bending	Suitable for Sustained Folding/Bending	Freq. (GHz)		Gain (dBi)		Eff. (%)		BW (%)		Dimension ($\lambda_{min} \times \lambda_{min}$)	Height (λ_{min})
				ε_r	$tan\ \delta$				Min.	Max.	Min.	Max.	Min.	Max.	Min.	Max.		
Origami Based antenna-physical deformation	[37]	E-threads	Threads	-	-	1.9 Ω/m	Yes	Yes	0.76	1.015	−2.5	0	N/A	N/A	12	14	0.42 × 0.025	N/A
		Organza	Fabric	(~1)	-	-												
	[74]	Copper film	Metal	-		4.4×10^5 S/m	Yes	No	0.9	2.5	1.1	3.28	N/A	N/A	14	20	0.15 × 0.15	0.0006 to 0.15
		Paper	Paper	2.2	0.04	-												
	[75]	Copper sheets	Metal	-	-	-	Yes	Yes	0.95	1.6	N/A	N/A	N/A	N/A	N/A	N/A	0.44 × 0.12	N/A
		Verowhite	Polymer	2.8	0.01	-												
	[92]	Copper	Metal	-	-	5.96×10^5 S/m	Yes	Yes	2.4		4.5	11	86	98	16	30	2.2 × 2.2 to 0.73 × 0.73	~0 to 0.73
		FR4 substrate	Composite	4.4	-	-												

In liquid metal-based antennas, power handling is more efficient and highly linear because electric switches are not used. Therefore, this technique is recommended for high-power applications. The antennas based on this technique are more flexible and durable compared to other techniques, due to the use of soft materials. In addition, their efficiency is higher compared to antennas based on electrical switches due to higher conductivity and lower dielectric loss of the materials. On the other hand, there are more limitations in the range of materials that can be used. It is very challenging to design a complex structure with this technique especially compared to their rigid counterparts. In an antenna which uses conductive fluid as a switch, the reconfigurability is mostly limited to two states.

Origami based antennas as a flexible structures do not use electrical switches. They offer a complexity reduction due to their method of reconfigurability based on changing the antenna's shape. The efficiency of these antennas is higher compare to textile or fabric-based antennas with electrical switches. This is primarily due to employment of copper tape, and other solid conductors in their structures. As they primarily use paper or copper in the form of thin foil in their structure, they are not robust or durable for continuous deformation compares to the other presented methods. In addition, there is a challenge of holding the antenna's shape upon folding and unfolding, therefore in some cases, an actuator is employed in the antenna structure. Furthermore, in some types of origami-based structure such as a magic cube, the reconfigurability is mostly limited to two states.

7. Conclusions

We critically reviewed the state-of-the-art in flexible and conformal reconfigurable antennas, categorizing of the developments with respect to frequency, pattern, polarization reconfigurability, as well as a combination of these features. Performance requirements and realization challenges that are unique to flexible and conformal reconfigurable antennas are discussed. Materials utilized in development of flexible and conformal reconfigurable antennas were considered critically, taking into account the applicability of these materials in niche applications, in the light of their electrical and mechanical characteristics. With an ever-increasing interest in flexible and conformal reconfigurable antennas nowadays, we conclude with identifying the following open research topics and potential directions for future research:

- Characterization of new flexible materials with low-moderate loss at mm-wave and terahertz frequencies;
- Robust integration methods to combine switching elements with flexible materials;
- Development if switch-less reconfiguration approaches;
- Integrating MEMS switches on flexible materials to realize this type of antenna;
- Integrating optical switches on flexible materials to realize this type of antenna;
- Employing reconfigurable microfluidic antennas for complex patch structures.

Author Contributions: Conceptualization, B.M. and R.M.H.; methodology, B.M.; investigation, B.M.; resources, B.M. and R.B.V.B.S.; data curation, B.M.; writing—original draft preparation, B.M.; writing—review and editing, R.B.V.B.S. and R.M.H. and S.M., A.L.; supervision, R.M.H. and K.P.E.; project administration, R.M.H. and K.P.E.; funding acquisition, B.M. and K.P.E. All authors have read and agreed to the published version of the manuscript.

Funding: This work was supported in part by the International Macquarie University Research Excellence Scholarship.

Acknowledgments: The authors would like thank the editors of this special issue for inviting us to submit a manuscript on this topic.

Conflicts of Interest: The authors declare no conflict of interest.

References

1. Salleh, S.M.; Jusoh, M.; Ismail, A.H.; Kamarudin, M.R.; Nobles, P.; Rahim, M.K.A.; Sabapathy, T.; Osman, M.N.; Jais, M.I.; Soh, P.J. Textile antenna with simultaneous frequency and polarization reconfiguration for WBAN. *IEEE Access* **2018**, *6*, 7350–7358. [CrossRef]

2. Haider, N.; Caratelli, D.; Yarovoy, A. Recent developments in reconfigurable and multiband antenna technology. *Int. J. Antennas Propag.* **2013**, *2013*, 869170. [CrossRef]

3. Sayem, A.S.M.; Simorangkir, R.B.V.B.; Esselle, K.P.; Hashmi, R.M.; Liu, H. A Method to Develop Flexible Robust Optically Transparent Unidirectional Antennas Utilizing Pure Water, PDMS and Transparent Conductive Mesh. *IEEE Trans. Antennas Propag.* **2020**. [CrossRef]

4. Sayem, A.S.M.; Simorangkir, R.B.V.B.; Esselle, K.P.; Hashmi, R.M. Development of Robust Transparent Conformal Antennas Based on Conductive Mesh-Polymer Composite for Unobtrusive Wearable Applications. *IEEE Trans. Antennas Propag.* **2019**, *67*, 7216–7224. [CrossRef]

5. Anagnostou, D.; Iskander, M. Adaptive Flexible Antenna Array System for Deformable Wing Surfaces. In Proceedings of the 2015 IEEE Aerospace Conference, Big Sky, MT, USA, 7–14 March 2015; pp. 1–6.

6. Anagnostou, D.E. Flexible arrays signal change in communications. *Nat. Electron.* **2019**, *2*, 180–181. [CrossRef]

7. Varnoosfaderani, M.V.; Thiel, D.V.; Lu, J. External parasitic elements on clothing for improved performance of wearable antennas. *IEEE Sens. J.* **2014**, *15*, 307–315. [CrossRef]

8. Varnoosfaderani, M.V.; Thiel, D.V.; Lu, J.W. A wideband slot antenna in a box for wearable sensor nodes. *IEEE Antennas Wirel. Propag. Lett.* **2015**, *14*, 1494–1497. [CrossRef]

9. Parchin, N.O.; Basherlou, H.J.; Al-Yasir, Y.I.; Abdulkhaleq, A.M.; Abd-Alhameed, R.A. Reconfigurable antennas: Switching techniques—A survey. *Electronics* **2020**, *9*, 336. [CrossRef]

10. Motovilova, E.; Huang, S.Y. A review on reconfigurable liquid dielectric antennas. *Materials* **2020**, *13*, 1863. [CrossRef]

11. Rutschlin, V.; Sokol, V. Reconfigurable antenna simulation. *IEEE Microw. Mag.* **2013**, *14*, 92–101. [CrossRef]

12. Oliveri, G.; Werner, D.H.; Massa, A. Reconfigurable electromagnetics through metamaterials—A review. *Proc. IEEE* **2015**, *103*, 1034–1056. [CrossRef]

13. Petosa, A. An overview of tuning techniques for frequency-agile antennas. *IEEE Antennas Propag. Mag.* **2012**, *54*, 271–296. [CrossRef]

14. Parchin, N.O.; Basherlou, H.J.; Al-Yasir, Y.I.; Abd-Alhameed, R.A.; Abdulkhaleq, A.M.; Noras, J.M. Recent developments of reconfigurable antennas for current and future wireless communication systems. *Electronics* **2019**, *8*, 128. [CrossRef]

15. Christodoulou, C.G.; Tawk, Y.; Lane, S.A.; Erwin, S.R. Reconfigurable antennas for wireless and space applications. *Proc. IEEE* **2012**, *100*, 2250–2261. [CrossRef]

16. Haupt, R.L.; Lanagan, M. Reconfigurable antennas. *IEEE Antennas Propag. Mag.* **2013**, *55*, 49–61. [CrossRef]

17. Kumar, J.; Basu, B.; Talukdar, F.A.; Nandi, A. Graphene-based multimode inspired frequency reconfigurable user terminal antenna for satellite communication. *IET Commun.* **2017**, *12*, 67–74. [CrossRef]

18. Javed, A.; Bhellar, B.; Tahir, F.A. Reconfigurable Body Worn Antenna for Bluetooth and WiMAX. In Proceedings of the 12th International Bhurban Conference on Applied Sciences and Technology (IBCAST), Islamabad, Pakistan, 13–17 January 2015; pp. 571–573.

19. Khan, M.; Shah, S.; Ullah, S. *Dual-Band Frequency Reconfigurable Microstrip Patch Antenna on Wearable Substrate for WiFi and WiMAX Applications*; Technical Journal; University of Engineering and Technology: Taxila, Pakistan, 2017; Volume 22, pp. 35–40.

20. Tahir, F.A.; Javed, A. A compact dual-band frequency reconfigurable textile antenna for wearable applications. *Microw. Opt. Technol. Lett.* **2015**, *57*, 2251–2257. [CrossRef]

21. Saraswat, K.; Harish, A.R. Flexible dual-band dual-polarised CPW-fed monopole antenna with discrete-frequency reconfigurability. *IET Microw. Antennas Propag.* **2019**, *13*, 2053–2060. [CrossRef]

22. Saeed, S.M.; Balanis, C.A.; Birtcher, C.R. Inkjet-printed flexible reconfigurable antenna for conformal WLAN/WiMAX wireless devices. *IEEE Antennas Wirel. Propag. Lett.* **2016**, *15*, 1979–1982. [CrossRef]

23. Sabapathy, T.; Bashah, M.; Jusoh, M.; Soh, P.; Kamarudin, M. Frequency Reconfigurable Rectangular Antenna with T-Slotted Feed Line.In Proceedings of the2016 International Conference on Radar, Antenna, Microwave, Electronics, and Telecommunications (ICRAMET), Jakarta, Indonesia, 3–5 October 2016; pp. 81–84.

24. Simorangkir, R.B.; Yang, Y.; Esselle, K.P.; Zeb, B.A. A method to realize robust flexible electronically tunable antennas using polymer-embedded conductive fabric. *IEEE Trans. Antennas Propag.* **2017**, *66*, 50–58. [CrossRef]

25. Chen, S.J.; Ranasinghe, D.C.; Fumeaux, C. A robust Snap-On button solution for reconfigurable wearable textile antennas. *IEEE Trans. Antennas Propag.* **2018**, *66*, 4541–4551. [CrossRef]

26. Kasirga, T.S.; Ertas, Y.N.; Bayindir, M. Microfluidics for reconfigurable electromagnetic metamaterials. *Appl. Phys. Lett.* **2009**, *95*, 214102. [CrossRef]

27. King, A.J.; Patrick, J.F.; Sottos, N.R.; White, S.R.; Huff, G.H.; Bernhard, J.T. Microfluidically switched frequency reconfigurable slot antennas. *IEEE Antennas Wirel. Propag. Lett.* **2013**, *12*, 828–831. [CrossRef]

28. Murray, C.; Franklin, R.R. Frequency Tunable Fluidic Annular Slot Antenna.In Proceedings of the2013 IEEE Antennas and Propagation Society International Symposium (APSURSI), Orlando, FL, USA, 7–13 July 2013; pp. 386–387.

29. Murray, C.; Franklin, R.R. Independently tunable annular slot antenna resonant frequencies using fluids. *IEEE Antennas Wirel. Propag. Lett.* **2014**, *13*, 1449–1452. [CrossRef]

30. Chaudhari, S.; Alharbi, S.; Zou, C.; Shah, H.; Harne, R.L.; Kiourti, A. A New Class of Reconfigurable Origami Antennas Based on E-Textile Embroidery.In Proceedings of the2018 IEEE International Symposium on Antennas and Propagation & USNC/URSI National Radio Science Meeting, Boston, MA, USA, 8–13 July 2018; pp. 183–184.

31. Liu, X.; Yao, S.; Cook, B.S.; Tentzeris, M.M.; Georgakopoulos, S.V. An origami reconfigurable axial-mode bifilar helical antenna. *IEEE Trans. Antennas Propag.* **2015**, *63*, 5897–5903. [CrossRef]

32. Hayes, G.J.; Liu, Y.; Genzer, J.; Lazzi, G.; Dickey, M.D. Self-folding origami microstrip antennas. *IEEE Trans. Antennas Propag.* **2014**, *62*, 5416–5419. [CrossRef]

33. Yao, S.; Liu, X.; Georgakopoulos, S.V. Morphing origami conical spiral antenna based on the Nojima wrap. *IEEE Trans. Antennas Propag.* **2017**, *65*, 2222–2232. [CrossRef]

34. Mazlouman, S.J.; Mahanfar, A.; Menon, C.; Vaughan, R.G. Reconfigurable axial-mode helix antennas using shape memory alloys. *IEEE Trans. Antennas Propag.* **2011**, *59*, 1070–1077. [CrossRef]

35. Liu, X.; Yao, S.; Georgakopoulos, S.V. A Frequency Tunable Origami Spherical Helical Antenna.In Proceedings of the2017 IEEE International Symposium on Antennas and Propagation & USNC/URSI National Radio Science Meeting, San Diego, CA, USA, 9–14 July 2017; pp. 1361–1362.

36. Yao, S.; Liu, X.; Gibson, J.; Georgakopoulos, S.V. Deployable Origami Yagi Loop Antenna.In Proceedings of the2015 IEEE International Symposium on Antennas and Propagation & USNC/URSI National Radio Science Meeting, Vancouver, BC, Canada, 19–24 July 2015; pp. 2215–2216.

37. Alharbi, S.; Chaudhari, S.; Inshaar, A.; Shah, H.; Zou, C.; Harne, R.L.; Kiourti, A. E-textile origami dipole antennas with graded embroidery for adaptive RF performance. *IEEE Antennas Wirel. Propag. Lett.* **2018**, *17*, 2218–2222. [CrossRef]

38. Lee, S.; Shah, S.I.H.; Lee, H.L.; Lim, S. Frequency reconfigurable antenna inspired by origami flasher. *IEEE Antennas Wirel. Propag. Lett.* **2019**, *18*, 1691–1695. [CrossRef]

39. Khan, M.R.; Zekios, C.L.; Bhardwaj, S.; Georgakopoulos, S.V. Origami-Enabled Frequency Reconfigurable Dipole Antenna.In Proceedings of the2019 IEEE International Symposium on Antennas and Propagation and USNC-URSI Radio Science Meeting, Atlanta, GA, USA, 7–12 July 2019; pp. 901–902.

40. Shah, S.I.H.; Lim, S. Frequency switchable origami magic cube antenna.In Proceedings of the2017 IEEE Asia Pacific Microwave Conference (APMC), Kuala Lumpar, Malaysia, 13–16 November 2017; pp. 105–107.

41. Zainud-Deen, S.H.; Malhat, H.A.E.-A.; El-Shalaby, N.A.A.S.; Gaber, S.M. Circular polarization bandwidth reconfigurable high gain planar plasma helical antenna. *IEEE Trans. Plasma Sci.* **2019**, *47*, 4274–4280. [CrossRef]

42. Liu, X.; Yao, S.; Georgakopoulos, S.V.; Tentzeris, M. Origami Quadrifilar Helix Antenna in UHF Band. In Proceedings of the 2014 IEEE Antennas and Propagation Society International Symposium (APSURSI), Memphis, TN, USA, 6–11 July 2014; pp. 372–373.

43. Liu, X.; Georgakopoulos, S.V.; Tentzeris, M. A Novel Mode and Frequency Reconfigurable Origami Quadrifilar Helical Antenna.In Proceedings of the2015 IEEE 16th Annual Wireless and Microwave Technology Conference (WAMICON), Cocoa Beach, FL, USA, 13–15 April 2015; pp. 1–3.

44. Liu, X.; Yao, S.; Georgakopoulos, S.V. Frequency Reconfigurable Origami Quadrifilar Helical Antenna with Reconfigurable Reflector.In Proceedings of the2015 IEEE International Symposium on Antennas and Propagation & USNC/URSI National Radio Science Meeting, Vancouver, BC, Canada, 19–24 July 2015; pp. 2217–2218.

45. Liu, X.; Yao, S.; Gonzalez, P.; Georgakopoulos, S.V. A Novel Ultra-Wideband Origami Reconfigurable Quasi-Taper Helical Antenna.In Proceedings of the2016 IEEE International Symposium on Antennas and Propagation (APSURSI), Fajardo, Puerto Rico, 26 June–1 July 2016; pp. 839–840.

46. Liu, X.; Yao, S.; Russo, N.; Georgakopoulos, S.V. Reconfigurable helical antenna based on origami neoprene with high radiation efficiency.In Proceedings of the2018 IEEE International Symposium on Antennas and Propagation & USNC/URSI National Radio Science Meeting, Boston, MA, USA, 8–13 July 2018; pp. 185–186.

47. Liu, X.; Yao, S.; Russo, N.; Georgakopoulos, S. Tri-band Reconfigurable Origami Helical Array.In Proceedings of the2018 IEEE International Symposium on Antennas and Propagation & USNC/URSI National Radio Science Meeting, Boston, MA, USA, 8–13 July 2018; pp. 1231–1232.

48. Carrara, G.P.; Russo, N.E.; Zekios, C.L.; Georgakopoulos, S.V. A Deployable and Reconfigurable Origami Antenna for extended Mobile Range.In Proceedings of the2019 IEEE International Symposium on Antennas and Propagation and USNC-URSI Radio Science Meeting, Atlanta, GA, USA, 7–12 July 2019; pp. 453–454.

49. Liu, X.; Yao, S.; Georgakopoulos, S.V.; Cook, B.S.; Tentzeris, M.M. Reconfigurable Helical Antenna Based on an Origami Structure for Wireless communication system.In Proceedings of the2014 IEEE MTT-S International Microwave Symposium (IMS2014), Tampa, FL, USA, 1–6 June 2014; pp. 1–4.

50. Yao, S.; Bao, K.; Liu, X.; Georgakopoulos, S.V. Tunable UHF Origami Spring Antenna with Actuation System.In Proceedings of the2017 IEEE International Symposium on Antennas and Propagation & USNC/URSI National Radio Science Meeting, San Diego, CA, USA, 9–14 July 2017; pp. 325–326.

51. Yao, S.; Georgakopoulos, S.V.; Cook, B.; Tentzeris, M. A Novel Reconfigurable Origami Accordion Antenna.In Proceedings of the2014 IEEE MTT-S International Microwave Symposium (IMS2014), Tampa, FL, USA, 1–6 June 2014; pp. 1–4.

52. Cardoso, J.; Safaai-Jazi, A. Spherical helical antenna with circular polarisation over a broad beam. *Electron. Lett.* **1993**, *29*, 325–326. [CrossRef]

53. Seiler, S.R.; Bazzan, G.; Fuchi, K.; Alanyak, E.J.; Gillman, A.S.; Reich, G.W.; Buskohl, P.R.; Pallampati, S.; Sessions, D.; Grayson, D.; et al. Physical Reconfiguration of an origami-inspired deployable Microstrip Patch Antenna Array.In Proceedings of the2017 IEEE International Symposium on Antennas and Propagation & USNC/URSI National Radio Science Meeting, San Diego, CA, USA, 9–14 July 2017; pp. 2359–2360.

54. Saeed, S.M.; Balanis, C.A.; Birtcher, C.R.; Durgun, A.C.; Shaman, H.N. Wearable flexible reconfigurable antenna integrated with artificial magnetic conductor. *IEEE Antennas Wirel. Propag. Lett.* **2017**, *16*, 2396–2399. [CrossRef]

55. Mohamadzade, B.; Hashmi, R.M.; Simorangkir, R.B.; Gharaei, R.; Rehman, S.U.; Abbasi, Q.H. Recent advances in fabrication methods for flexible antennas in wearable devices: State of the art. *Sensors* **2019**, *19*, 2312. [CrossRef]

56. Mohamadzade, B.; Simorangkir, R.B.; Hashmi, R.M.; Lalbakhsh, A. A conformal ultrawideband antenna with monopole-like radiation patterns. *IEEE Trans. Antennas Propag.* **2020**, *68*, 6383–6388. [CrossRef]

57. Mohamadzade, B.; Simorangkir, R.B.; Hashmi, R.M.; ChaoOger, Y.; Zhadobov, M.; Sauleau, R. A conformal band-notched ultrawideband antenna with monopole-like radiation characteristics. *IEEE Antennas Wirel. Propag. Lett.* **2019**, *19*, 203–207. [CrossRef]

58. Shakhirul, M.; Jusoh, M.; Ismail, A.; Kamarudin, M.; Rahim, H.A.; Sabapathy, T. Reconfigurable Frequency with Circular Polarization for on-Body Wearable Textile Antenna.In Proceedings of the2016 10th European Conference on Antennas and Propagation (EuCAP), Davos, Switzerland, 10–15 April 2016; pp. 1–4.

59. Dang, Q.H.; Chen, S.J.; Ranasinghe, D.C.; Fumeaux, C. A Reconfiguration Module with Coplanar Snap-On Connection for Wearable Textile Antennas.In Proceedings of the2019 IEEE Asia-Pacific Microwave Conference (APMC), Singapore, Singapore, 10–13 December 2019; pp. 613–615.

60. Kingsley, N.; Anagnostou, D.E.; Tentzeris, M.; Papapolymerou, J. RF MEMS sequentially reconfigurable Sierpinski antenna on a flexible organic substrate with novel dc-biasing technique. *J. Microelectromechan. Syst.* **2007**, *16*, 1185–1192. [CrossRef]

61. Bairavasubramanian, R.; Kingsley, N.; DeJean, G.; Wang, G.; Anagnostou, D.E.; Tentzeris, M.; Papapolymerou, J. Recent Developments on Lightweight, Flexible, Dual Polarization/Frequency Phased Arrays Using RF Mems Switches on LCP Multilayer Substrates for Remote Sensing of Precipitation. In Proceedings of the 6th Annual NASA Earth Science Technology Conference (ESTC 2006), College Park, MD, USA, 27–29 June 2006.

62. Anagnostou, D.E.; Bairavasubramanian, R.; DeJean, G.; Wang, G.; Kingsley, N.; Tentzeris, M.; Papapolymerou, J. Development of a Dual-Frequency, Dual-Polarization, Flexible and Deployable Antenna Array for Weather Applications.In Proceedings of the15th IST Mobile & Wireless Communication Summit, Myconos, Greece, 4–8 June 2006.

63. Jang, T.; Zhang, C.; Youn, H.; Zhou, J.; Guo, L.J. Semitransparent and flexible mechanically reconfigurable electrically small antennas based on tortuous metallic micromesh. *IEEE Trans. Antennas Propag.* **2016**, *65*, 150–158. [CrossRef]

64. Saghati, A.P.; Batra, J.S.; Kameoka, J.; Entesari, K. Miniature and reconfigurable CPW folded slot antennas employing liquid-metal capacitive loading. *IEEE Trans. Antennas Propag.* **2015**, *63*, 3798–3807. [CrossRef]

65. Saghati, A.P.; Batra, J.; Kameoka, J.; Entesari, K. A Microfluidically-Switched CPW Folded Slot Antenna.In Proceedings of the2014 IEEE Antennas and Propagation Society International Symposium (APSURSI), Memphis, TN, USA, 6–11 July 2014; pp. 557–558.

66. So, J.-H.; Thelen, J.; Qusba, A.; Hayes, G.J.; Lazzi, G.; Dickey, M.D. Reversibly deformable and mechanically tunable fluidic antennas. *Adv. Funct. Mater.* **2009**, *19*, 3632–3637. [CrossRef]

67. Mazlouman, S.J.; Jiang, X.J.; Mahanfar, A.; Menon, C.; Vaughan, R.G. A Reconfigurable Patch Antenna Using Liquid Metal Embedded in a Silicone Substrate. *IEEE Trans. Antennas Propag.* **2011**, *59*, 4406–4412. [CrossRef]

68. Zandvakili, M.; Honari, M.M.; Sameoto, D.; Mousavi, P. Microfluidic Liquid Metal Based Mechanically Reconfigurable Antenna Using Reversible Gecko Adhesive Based Bonding.In Proceedings of the2016 IEEE MTT-S International Microwave Symposium (IMS), San Francisco, CA, USA, 22–27 May 2016; pp. 1–4.

69. Su, W.; Nauroze, S.A.; Ryan, B.; Tentzeris, M.M. Novel 3d Printed Liquid-Metal-Alloy Microfluidics-Based Zigzag and Helical Antennas for Origami Reconfigurable Antenna "Trees".In Proceedings of the2017 IEEE MTT-S International Microwave Symposium (IMS), Honolulu, HI, USA, 4–9 June 2017; pp. 1579–1582.

70. Georgakopoulos, S.V. Reconfigurable Origami Antennas.In Proceedings of the2019 International Applied Computational Electromagnetics Society Symposium (ACES), Miami, FL, USA, 14–19 April 2019; pp. 1–2.

71. Fuchi, K.; Buskohl, P.R.; Joo, J.J.; Reich, G.W.; Vaia, R.A. Resonance Tuning of RF Devices through Origami Folding.In Proceedings of the20th International Conference on Composite Materials, Copenhagen, Denmark, 19–24 July 2015; pp. 1–10.

72. Liu, X.; Georgakopoulos, S.V.; Rao, S. A design of an origami reconfigurable QHA with a foldable reflector [antenna applications corner]. *IEEE Antennas Propag. Mag.* **2017**, *59*, 78–105. [CrossRef]

73. Liu, X.; Zekios, C.L.; Georgakopoulos, S.V. Analysis of a packable and tunable origami multi-radii helical antenna. *IEEE Access* **2019**, *7*, 13003–13014. [CrossRef]

74. Shah, S.I.H.; Lim, S. A dual band frequency reconfigurable origami magic cube antenna for wireless sensor network applications. *Sensors* **2017**, *17*, 2675. [CrossRef]

75. Liu, C.-X.; Zhang, Y.-J.; Li, Y. Precisely manipulated origami for consecutive frequency reconfigurable antennas. *Smart Mater. Struct.* **2019**, *28*, 095020. [CrossRef]

76. Lee, S.; Lee, M.; Lim, S. Frequency reconfigurable antenna actuated by three-storey tower kirigami. *Extrem. Mech. Lett.* **2020**, *39*, 100833. [CrossRef]

77. Nogi, M.; Komoda, N.; Otsuka, K.; Suganuma, K. Foldable nanopaper antennas for origami electronics. *Nanoscale* **2013**, *5*, 4395–4399. [CrossRef]

78. Jilani, S.F.; Rahimian, A.; Alfadhl, Y.; Alomainy, A. Low-profile flexible frequency-reconfigurable millimetre-wave antenna for 5G applications. *Flex. Print. Electron.* **2018**, *3*, 035003. [CrossRef]

79. Jilani, S.F.; Greinke, B.; Hao, Y.; Alomainy, A. Flexible Millimetre-Wave Frequency Reconfigurable Antenna for Wearable Applications in 5G Networks.In Proceedings of the2016 URSI International Symposium on Electromagnetic Theory (EMTS), Espoo, Finland, 14–18 August 2016; pp. 846–848.

80. Radha, S.; Rao, P.H. Pattern Reconfigurable Ultra Wideband Thin Antenna.In Proceedings of the2012 Third International Conference on Computing, Communication and Networking Technologies (ICCCNT'12), Coimbatore, India, 26–28 July 2012; pp. 1–5.

81. Jais, M.; Jamlos, M.; Jusoh, M.; Sabapathy, T.; Kamarudin, M. 2.45 GHz Beam-Steering Textile Antenna for WBAN Application.In Proceedings of the2013 IEEE Antennas and Propagation Society International Symposium (APSURSI), Orlando, FL, USA, 7–13 July 2013; pp. 200–201.

82. He, X.; Gao, P.; Zhu, Z.; You, S.; Wang, P. A flexible pattern reconfigurable antenna for WLAN wireless systems. *J. Electromagn. Waves Appl.* **2019**, *33*, 782–793. [CrossRef]

83. Mohamadzade, B.; Simorangkir, R.B.; Hashmi, R.M.; Shrestha, S. Low-Profile Pattern Reconfigurable Antenna for Wireless Body Area Networks.In Proceedings of the2019 International Conference on Electromagnetics in Advanced Applications (ICEAA), Granada, Spain, 9–13 September 2019; pp. 546–547.

84. Mohamadzade, B.; Simorangkir, R.B.; Hashmi, R.M.; Gharaei, R.; Lalbakhsh, A. Monopole-like and semi-directional reconfigurable pattern antenna for wireless body area network applications. *Microw. Opt. Technol. Lett.* **2019**, *61*, 2760–2765. [CrossRef]

85. Ha, S.-J.; Jung, Y.-B.; Kim, Y.; Jung, C.W. Reconfigurable beam-steering antenna using dipole and loop combined structure for wearable applications. *ETRI J.* **2012**, *34*, 1–8. [CrossRef]

86. Yan, S.; Vandenbosch, G.A. Radiation pattern-reconfigurable wearable antenna based on metamaterial structure. *IEEE Antennas Wirel. Propag. Lett.* **2016**, *15*, 1715–1718. [CrossRef]

87. Yan, S.; Vandenbosch, G.A. Wearable Pattern Reconfigurable Patch Antenna.In Proceedings of the2016 IEEE International Symposium on Antennas and Propagation (APSURSI), Puerto Rico, USA, 26 June–1 July 2016; pp. 1665–1666.

88. Mendes, C.; Peixeiro, C. A dual-mode single-band wearable microstrip antenna for body area networks. *IEEE Antennas Wirel. Propag. Lett.* **2017**, *16*, 3055–3058. [CrossRef]

89. Di Natale, A.; Di Giampaolo, E. A reconfigurable all-textile wearable UWB antenna. *Prog. Electromagn. Res.* **2020**, *103*, 31–43. [CrossRef]

90. Moghadas, H.; Zandvakili, M.; Sameoto, D.; Mousavi, P. Beam reconfigurable aperture antenna by stretching or reshaping of a flexible surface. *IEEE Antennas Wirel. Propag. Lett.* **2016**, *16*, 1337–1340. [CrossRef]

91. Shah, S.I.H.; Ghosh, S.; Lim, S. A Novel Bio Inspired Pattern Reconfigurable Quasi-Yagi Helical Antenna using Origami DNA.In Proceedings of the2018 International Symposium on Antennas and Propagation (ISAP), Busan, Korea, 23–26 October 2018.

92. Shah, S.I.H.; Lim, S. Thermally beam-direction-and beamwidth switchable monopole antenna using origami reflectors with smart shape memory polymer hinges. *IEEE Antennas Wirel. Propag. Lett.* **2019**, *18*, 1696–1700. [CrossRef]

93. Shah, S.I.H.; Tentzeris, M.M.; Lim, S. A deployable quasiyagi monopole antenna using three origami magic spiral cubes. *IEEE Antennas Wirel. Propag. Lett.* **2018**, *18*, 147–151. [CrossRef]

94. Hatsopoulos, N.G.; Warren, W.H., Jr. Resonance tuning in rhythmic arm movements. *J. Mot. Behav.* **1996**, *28*, 3–14. [CrossRef]

95. Xu, Y.; Kim, Y.; Tentzeris, M.M.; Lim, S. Bi-directional loop antenna array using magic cube origami. *Sensors* **2019**, *19*, 3911. [CrossRef]

96. Jais, M.I.; Jamlos, M.; Jusoh, M.; Sabapathy, T.; Kamarudin, M.R. A novel 1.575-GHz dual-polarization textile antenna for gps application. *Microw. Opt. Technol. Lett.* **2013**, *55*, 2414–2420. [CrossRef]

97. Lee, C.M.; Jung, C.W. Radiation-pattern-reconfigurable antenna using monopole-loop for Fitbit flex wristband. *IEEE Antennas Wirel. Propag. Lett.* **2014**, *14*, 269–272. [CrossRef]

98. Liu, X.; Yao, S.; Georgakopoulos, S.V. Mode Reconfigurable Bistable Spiral Antenna Based on Kresling Origami.In Proceedings of the2017 IEEE International Symposium on Antennas and Propagation & USNC/URSI National Radio Science Meeting, San Diego, CA, USA, 9–14 July 2017; pp. 413–414.

99. Bhattacharjee, S.; Teja, S.; Midya, M.; Chaudhuri, S.B.; Mitra, M. Dual Band Dual Mode Triangular Textile Antenna for Body-Centric Communications.In Proceedings of the2019 URSI Asia-Pacific Radio Science Conference (AP-RASC), New Delhi, India, 9–15 March 2019; pp. 1–4.

100. Langley, R.J.; Ford, K.L.; Lee, H.-J. Switchable on/off-Body Communication at 2.45 GHz Using Textile Microstrip Patch Antenna on Stripline.In Proceedings of the2012 6th European Conference on Antennas and Propagation (EUCAP), Prague, Czech Republic, 26–30 March 2012; pp. 728–731.

101. Lee, H.; Choi, J. A Polarization Reconfigurable Textile Patch Antenna for Wearable IOT Applications. In Proceedings of the 2017 International Symposium on Antennas and Propagation (ISAP), Phuket, Thailand, 30 October–2 November 2017; pp. 1–2.

102. Jais, M.; Jamlos, M.; Jusoh, M.; Sabapathy, T.; Nayan, M.; Shahadah, A.; Rahim, H.A.; Ismail, I.; Fuad, F. 1.575 GHz Dual-Polarization Textile Antenna (DPTA) for GPS Application.In Proceedings of the2013 IEEE Symposium on Wireless Technology & Applications (ISWTA), Kuching, Malaysia, 22–25 September 2013; pp. 376–379.

103. Yao, S.; Georgakopoulos, S.V. Origami segmented helical antenna with switchable sense of polarization. *IEEE Access* **2017**, *6*, 4528–4536. [CrossRef]

104. Yao, S.; Liu, X.; Georgakopoulos, S.V.; Schamp, R. Polarization Switchable Origami Helical Antenna. In Proceedings of the 2016 IEEE International Symposium on Antennas and Propagation (APSURSI), Fajardo, Puerto Rico, 26 June–1 July 2016; pp. 1667–1668.

105. Yao, S.; Liu, X.; Georgakopoulos, S.V. Segmented Helical Antenna with Reconfigurable Polarization.In Proceedings of the2017 IEEE International Symposium on Antennas and Propagation & USNC/URSI National Radio Science Meeting, San Diego, CA, USA, 9–14 July 2017; pp. 2207–2208.
106. Tsolis, A.; Michalopoulou, A.; Alexandridis, A.A. Use of conductive zip and velcro as a polarisation reconfiguration means of a textile patch antenna. *IET Microw. Antennas Propag.* **2020**, *14*, 684–693. [CrossRef]
107. Liu, X.; Yao, S.; Georgakopoulos, S.V. Reconfigurable Origami Equiangular Conical Spiral Antenna.In Proceedings of the2015 IEEE International Symposium on Antennas and Propagation & USNC/URSI National Radio Science Meeting, Vancouver, BC, Canada, 19–24 July 2015; pp. 2263–2264.
108. Vasina, P.; Lacik, J. Textile Linear Polarization Reconfigurable Ring Slot Antenna for 5.8 GHz Band.In Proceedings of the2017 Conference on Microwave Techniques (COMITE), Brno, Czech Republic, 20–21 April 2017; pp. 1–4.
109. Saha, P.; Mitra, D.; Parui, S.K.A. frequency and polarization agile disc monopole wearable antenna for medical applications. *Radioengineering* **2020**, *29*, 75. [CrossRef]
110. Chen, S.J.; Ranasinghe, D.C.; Fumeaux, C. A Polarization/Frequency Interchangeable Patch for a Modular Wearable Textile Antenna.In Proceedings of the2017 11th European Conference on Antennas and Propagation (EUCAP), Paris, France, 19–24 March 2017; pp. 2483–2486.
111. Zhu, Z.; Wang, P.; You, S.; Gao, P. A flexible frequency and pattern reconfigurable antenna for wireless systems. *Prog. Electromagn. Res.* **2018**, *76*, 63–70.
112. da Conceição Andrade, A.; Fonseca, I.P.; Jilani, S.F.; Alomainy, A. Reconfigurable Textile-Based Ultra-Wideband Antenna for Wearable Applications.In Proceedings of the2016 10th European Conference on Antennas and Propagation (EuCAP), Davos, Switzerland, 10–15 April 2016; pp. 1–4.
113. Gibson, J.S.; Liu, X.; Georgakopoulos, S.V.; Wie, J.J.; Ware, T.H.; White, T.J. Reconfigurable antennas based on self-morphing liquid crystalline elastomers. *IEEE Access* **2016**, *4*, 2340–2348. [CrossRef]
114. Nojima, T. Origami Modeling of Functional Structures Based on Organic Patterns. Mater's Thesis, Kyoto University, Kyoto, Japan, 2002.
115. Yao, S.; Liu, X.; Georgakopoulos, S.V. A Mode Reconfigurable Nojima Origami Antenna.In Proceedings of the2015 IEEE International Symposium on Antennas and Propagation & USNC/URSI National Radio Science Meeting, Vancouver, BC, Canada, 19–25 July 2015; pp. 2237–2238.

Article

Reconfigurable Multiband Operation for Wireless Devices Embedding Antenna Boosters

Jaume Anguera [1,2,*], Aurora Andújar [1], José Luis Leiva [1], Oriol Massó [2], Joakim Tonnesen [3], Endre Rindalsholt [3], Rune Brandsegg [3] and Roberto Gaddi [4]

[1] R&D Department, Ignion, 08190 Barcelona, Spain; aurora.andujar@ignion.io (A.A.); joseluis.leiva@ignion.io (J.L.L.)

[2] Remote-IoT GRIT Research Group, Universitat Ramon LLull, 08019 Barcelona, Spain; jaume.anguera@salle.url.edu

[3] Nordic Semiconductor, 7052 Trondheim, Norway; joakim.tonnesen@nordicsemi.no (J.T.); endre.rindalsholt@nordicsemi.no (E.R.); rune.brandsegg@nordicsemi.no (R.B.)

[4] Qorvo BV, 3511 SB Utrecht, The Netherlands; roberto.gaddi@qorvo.com

* Correspondence: jaume.anguera@ignion.io

Citation: Anguera, J.; Andújar, A.; Leiva, J.L.; Massó, O.; Tonnesen, J.; Rindalsholt, E.; Brandsegg, R.; Gaddi, R. Reconfigurable Multiband Operation for Wireless Devices Embedding Antenna Boosters. *Electronics* **2021**, *10*, 808. https://doi.org/10.3390/electronics10070808

Academic Editors: Dimitris Anagnostou, Michael Chryssomallis and Sotirios Goudos

Received: 31 December 2020
Accepted: 3 March 2021
Published: 29 March 2021

Publisher's Note: MDPI stays neutral with regard to jurisdictional claims in published maps and institutional affiliations.

Abstract: Wireless devices such as smart meters, trackers, and sensors need connections at multiple frequency bands with low power consumption, thus requiring multiband and efficient antenna systems. At the same time, antennas should be small to easily fit in the scarce space existing in wireless devices. Small, multiband, and efficient operation is addressed here with non-resonant antenna elements, featuring volumes less than 90 mm^3 for operating at 698–960 MHz as well as some bands in a higher frequency range of 1710–2690 MHz. These antenna elements are called antenna boosters, since they excite currents on the ground plane of the wireless device and do not rely on shaping complex geometric shapes to obtain multiband behavior, but rather the design of a multiband matching network. This design approach results in a simpler, easier, and faster method than creating a new antenna for every device. Since multiband operation is achieved through a matching network, frequency bands can be configured and optimized with a reconfigurable matching network. Two kinds of reconfigurable multiband architectures with antenna boosters are presented. The first one includes a digitally tunable capacitor, and the second one includes radiofrequency switches. The results show that antenna boosters with reconfigurable architectures feature multiband behavior with very small sizes, compared with other prior-art techniques.

Keywords: smart devices; reconfigurable antennas; antenna boosters; multiband antennas; small antennas; wireless devices; matching networks; digitally tunable capacitors; RF switching

1. Introduction

With the advent of the everything-connected era, antennas play a very significant role. From smart meters and smart trackers to sensors and any other Internet of Things (IoT) device, every wireless device needs at least one antenna to transmit and receive information. In response, the wireless industry is moving fast to launch new devices into this growing market. This fact pushes RF or microwave and wireless engineers to design radiofrequency chains in a smart way, using off-the-shelf devices such as filters, amplifiers, diplexers, and front-end modules. At the end of this radiofrequency chain, an antenna is needed to efficiently transmit and receive electromagnetic waves. Accordingly, an off-the-shelf antenna component would be an attractive selection for simplifying the entire design process. It would not only simplify the antenna design phase, but also the device manufacturing process, reducing costs and being compatible with pick-and-place machines for mass production.

Besides off-the-shelf antenna components, a miniature size and multiband performance are two other demanding requirements for wireless antenna systems. Common techniques to design small and multiband antennas for wireless devices rely on complex geometries, where the several resonant modes of the antenna determine the frequency bands of operation [1–11]. On the one hand, this procedure requires a high level of expertise to shape the antenna geometry correctly and achieve acceptable behavior to operate at the frequency bands of interest. This complexity of the antenna geometry shaping process makes RF or microwave and wireless engineers reluctant to design their own antennas for wireless systems. On the other hand, some currently available off-the-shelf antennas are mainly narrowband, while others, although multiband, are difficult to adjust to the bands of interest once they are integrated inside the device. The off-the-shelf antenna proposed here overcomes these shortcomings. A very small and simple element can be integrated inside any wireless device for operating at any frequency band through the proper adjustment of a matching network. The antenna element can be easily adjusted not by modifying its geometry, but through the proper matching network design. This is a simpler, faster, and more familiar method for RF or microwave and wireless engineers who are acquainted with matching networks at every single stage of a telecommunication system (e.g., filters and amplifiers). Therefore, designing a multiband antenna system with this technique is more related to a microwave circuit design than to an antenna geometry design, thus making antenna design much more like filter design than conventional antenna design.

The antenna technology relies on a non-resonant antenna element that is electrically short in terms of the operating wavelength (~$\lambda/30$). Its correct location on a ground plane of a wireless device enables the excitation of efficient radiating modes on the ground plane [12–17]. Therefore, the term antenna booster is adopted here to refer to this non-resonant antenna element. Antenna boosters are miniature, surface-mount, off-the-shelf components that replace conventional customized PIFA and printed antennas for mobile or wireless connectivity applications (Figure 1). However, since the antenna booster is electrically small, its impedance is mainly reactive, and thus it is poorly matched at most of the bands of operation in the frequency range from 0.6 GHz up to 10.6 GHz, where many communication systems are allocated, such as cellular and short-range wireless communications. This mismatch issue is overcome with the addition of a multiband matching network, resulting in a multiband antenna system with a small antenna element and with competitive efficiency values [12–17].

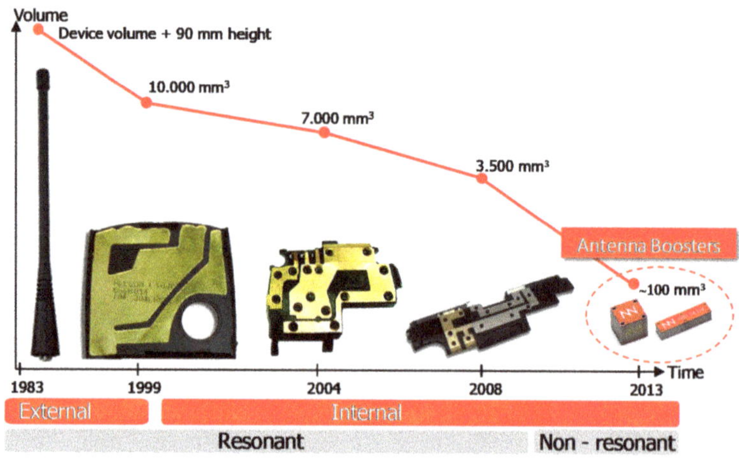

Figure 1. Progression from external, quarter wavelength monopole antennas to internal custom antennas based on complex geometries to the tiny, off-the-shelf antenna boosters.

Examples of antennas combined with matching networks are found in the literature [18–30]. However, all those designs rely on adding intelligence to the antenna geometry. In addition, in all the above cases, the antenna still represents a considerable volume. Although the volumes of the antennas have been continuously shrunk, they still rely on complex geometries (Figure 1). On the contrary, antenna boosters can still operate at multiple frequency bands while keeping a very small antenna volume [12–17].

In this paper, multiband operation is addressed with antenna boosters combined with reconfigurable matching networks, comprising lumped passive components (capacitors and inductors) as well as active devices. In a first case, a digitally tunable capacitor is shown, and in a second case, a switch-based solution is adopted. This results in reconfigurable matching networks, useful for tuning the frequency bands of operation. It is important to emphasize that a passive matching network can provide multiband operation. However, adding reconfigurability adds a new degree of freedom to expand the number of bands beyond the ones that can be achieved with a passive matching network. Moreover, as a second advantage, reconfigurability can advantageously be applied to retune the frequency bands of operation in wireless devices when exposed to different, harsh surroundings such as conductive bodies, concrete, wood, and plastics, which shift the frequency response of the antenna. Finally, the multiband behavior for small form factor devices (<50 mm × 50 mm) is challenged. Therefore, a reconfigurable antenna-booster architecture is convenient, because an antenna booster is small enough to be embedded into a small device, and reconfigurability provides connections at multiple frequency bands.

The paper is structured as follows. An introduction to antenna booster technology is explained in Section 2. Multiband reconfigurable matching networks using digitally tunable capacitors (DTCs) and RF switches are described in Sections 3 and 4, respectively. A comparison with prior-art multiband reconfigurable antennas is discussed in Section 5. Finally, conclusions are drawn in Section 6.

2. Antenna Booster Technology

Antenna booster-based technology relies on very small antenna elements called antenna boosters to provide operation in a wireless device (Figures 1 and 2)—[13]. The same antenna booster as the one depicted in Figure 2 can operate at different bands, depending on the matching network. This is a different approach than conventional antenna designs, where the antenna geometry must be created case by case, thus increasing the time and cost. In addition, the same antenna booster can operate for different platform sizes, since only the matching network changes from one platform to another [31]. This is a faster and easier approach to designing multiband antenna systems than designing an antenna from scratch.

Figure 2. An antenna booster (in red) located on a PCB (Printed Circuit Board) of a wireless device. Its tiny size (12 mm × 3 mm × 2.4 mm (height)) allows easy integration of the antenna booster within the device.

To illustrate the non-resonant nature of an antenna booster, an antenna booster (a 3D, brass-made parallelepiped in this example) of 12 mm × 3 mm × 2.4 mm (height) was analyzed, which represented a maximum size of only $\lambda/35.8$ at 698 MHz. The antenna booster was integrated in a ground plane featuring dimensions of 120 mm × 60 mm (Figure 3a). Said antenna booster was in charge of exciting radiating modes in the ground plane, which became the relevant part of the radiation process. This fact justifies the antenna booster name, since it boosted currents on the ground plane (Figure 4) [13]. From a practical perspective, the corners of the ground plane provided the best bandwidth and efficiency [17].

The input impedance was reactive, most particularly at the low bands (698–960 MHz), where the impedance was mainly capacitive (Figure 3b). A passive matching network could match the antenna system shown in (Figure 3c), covering 824–960 MHz up to 1710–2690 MHz, as shown in [17,32]. However, the frequency range from 698 MHz to 824 MHz would not be properly matched (Figure 3d). This frequency range is needed in many wireless devices, such as smart trackers. This frequency range represents 16% of the bandwidth, which supposes a significant challenge since it is located at the lower frequencies of operation. Therefore, in those situations where a passive matching network cannot squeeze more bandwidth, reconfigurability is a good option. In this sense, and with the aim of improving the performance in the frequency range of 698–960 MHz up to 1710–2200 MHz or 2690 MHz, reconfigurable matching architectures in combination with antenna boosters are proposed using digitally tunable capacitors (DTCs) and radio frequency switches in Sections 3 and 4, respectively.

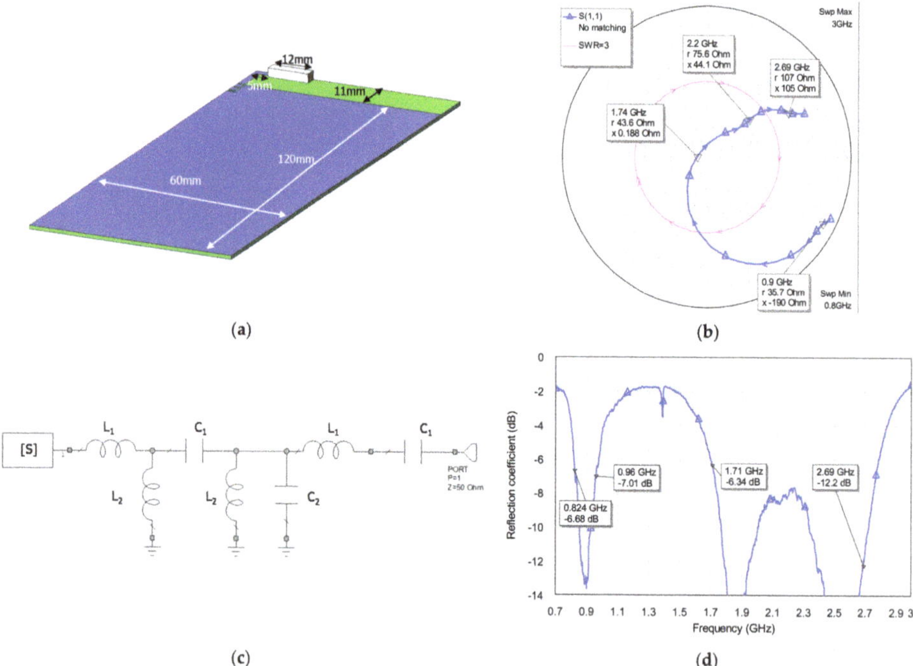

Figure 3. (**a**) Antenna booster on a 120 mm × 60 mm ground plane over a dielectric substrate ($\varepsilon_r = 4.15$, $\tan\delta = 0.013$, and h = 1 mm). The antenna booster is a 12 mm × 3 mm × 2.4 mm (height) 3D conductive parallelepiped. (**b**) Simulated input impedance, showing a reactive capacitive behavior at the low frequency region (698–960 MHz). (**c**) A passive matching network for achieving multiband behavior ($L_1 = 4.3$ nH, $L_2 = 18$ nH, $C_3 = 0.9$ pF, $C_4 = 1$ pF, $L_5 = 13$ nH, $C_6 = 2$ pF, and $L_7 = 4.5$ nH). (**d**) Reflection coefficient resulting from adding the matching network of (**c**) to the antenna system shown in (**a**).

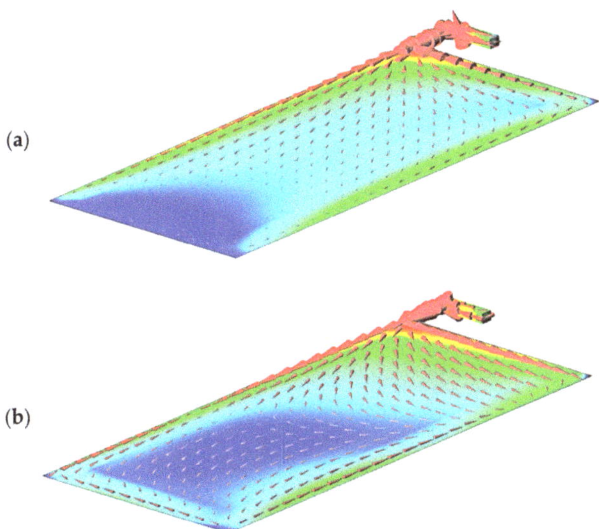

Figure 4. Simulated current distribution for the antenna booster shown in Figure 3a at (**a**) 850 MHz and (**b**) 2200 MHz. Most of the current is aligned with the long edge at 850 MHz, whereas the current at 2200 MHz has both directions: one aligned with the long edge and the other with the short edge. This current distribution depends on the ground plane size as well as the position of the antenna booster. Blue color represents the lowest intensity and red de maximum.

3. Digitally Tunable Capacitors for Antenna Boosters

The bandwidth of an antenna system cannot be infinitely enlarged by adding more components to a matching network, owing to the Fano limit [33]. To avoid such a drawback and increase the bandwidth to allocate more frequency bands, a tunable capacitor is strategically included in the passive matching network, resulting in a reconfigurable circuit. Hence, the reconfigurable circuit comprises lumped capacitors and inductors and a digitally tunable capacitor [34].

This section presents a multiband antenna system comprising an antenna booster and a reconfigurable matching network to cover 698–960 MHz and 1710–2690 MHz.

3.1. Digitally Tunable Capacitors

A digitally tunable capacitor (DTC) is a microwave device where the capacitance can be digitally changed within a range from C_{min} to C_{max}. The tuning ratio (C_{max}/C_{min}) is intrinsically related to the DTC quality factor (Q); the larger the tuning ratio, the lower the Q value is [35]. The microelectromechanical system (MEMS) technology DTC from Cavendish Kinetics (now Qorvo) for this design presented a value of Q = 700 at C_{max} = 1 pF at 700 MHz with a tuning ratio of 2.5. Thanks to the low intrinsic losses of MEMS technology, the DTC featured a high Q with a small footprint. In the present case, the DTC area was only 1.4 mm × 1.4 mm (Figure 5).

Besides 1.8 V (VIO, Figure 5), RF ground, and RF input connections, it supported the Mobile Industry Processor Interface (MIPI) radio frequency front-end (RFFE) interface with one clock (SCLK, Figure 5) and one series data (SDATA, Figure 5). With a 5-bit resolution in the digital control, there were 32 programmable capacitor states from 0.4 pF to 1.0 pF with a constant step size of 18 fF. The MIPI RFFE was compatible with other radiofrequency components and baseband processors in the system, which could select the state of the DTC based on the desired use condition. In the prototype described here, the state was selected using an MIPI master interface controlled by a laptop application. Once the state was set, the MIPI digital cable bus was disconnected in order to not interfere with the impedance and radiation measurements. An on-board battery and voltage regulator provided power to the DTC.

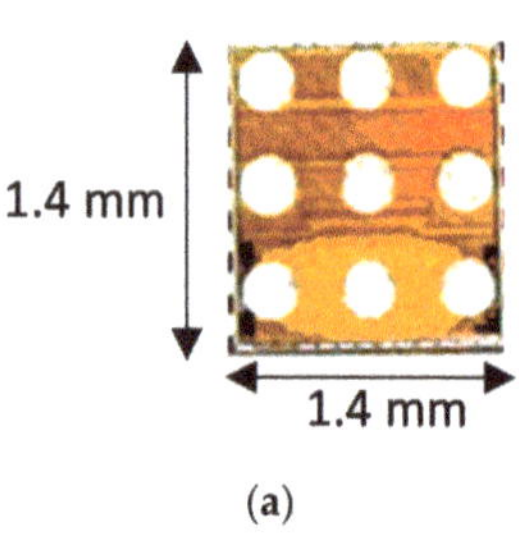

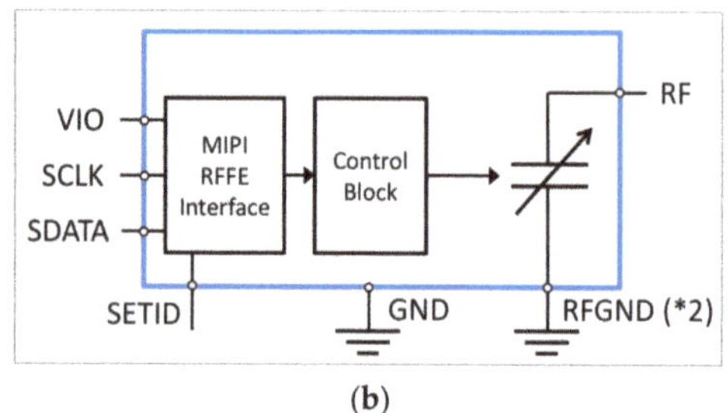

Figure 5. (**a**) Footprint of the digitally tunable capacitor (DTC). (**b**) Block diagram of the DTC, where the capacitor can be tuned from 0.4 pF to 1.0 pF. The equivalent series resistance (ESR) was ~0.3 Ω at C_{max} with a 5-bit resolution (32 capacitor states). The self-resonant frequency (SRF) was >15 GHz. The RF power handling was +39 dBm, and there was high linearity. For the Mobile Industry Processor Interface (MIPI) radio frequency front-end (RFFE) interface, the small size was ~2 mm^2, and the low power was a typical 180 µW.

3.2. A Reconfigurable Matching Network with a DTC

A small antenna booster (86.4 mm^3) was proposed for multiband operation in combination with a reconfigurable matching network [36]. The antenna booster was placed at the corner of a ground plane, featuring the dimensions (120 mm × 60 mm) of a wireless IoT device (Figure 6). Said reconfigurable matching network used the DTC implemented with MEMS described in the previous section. The RFFE digital signals for capacitance selection were provided through a removable ribbon cable. For the present case, only four states were enough to cover the frequency range of 698–960 MHz up to 1710–2690 MHz.

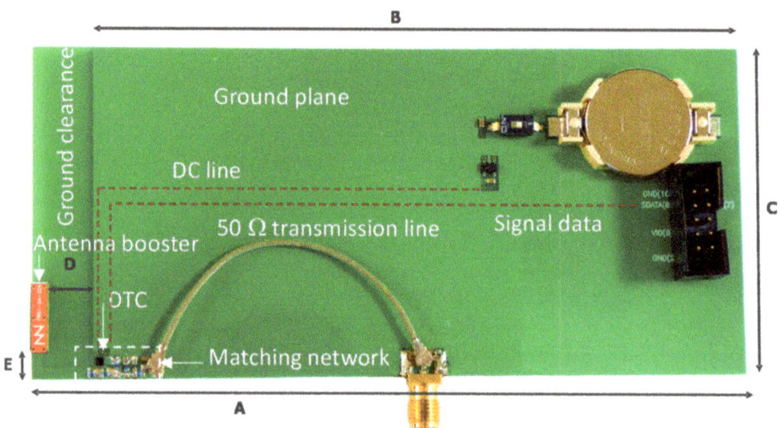

Figure 6. The antenna booster with the reconfigurable matching network and its control circuit. The control lines from the battery and RFFE interface to the tunable capacitor go underneath the ground plane (B × C) shown in the top layer of the board. A = 131 mm, B = 120 mm, C = 60 mm, E = 5 mm, and D = 8 mm.

Analysis was done using electromagnetic simulation of the PCB, including the antenna booster. The S-parameters computed with the electromagnetic simulation were then imported in a circuit simulator (cadence AWR Microwave Office) to design the matching network, comprising a combination of lumped capacitors, inductors, and the tunable capacitor. Once the simulation showed feasible results, a reference board was built (Figure 6).

The proposed reconfigurable matching network comprising a shunt DTC is shown in Figure 7. The programmable states for the DTC were S00 (0.40 pF), S02 (0.44 pF), S08 (0.55 pF), and S27 (0.92 pF).

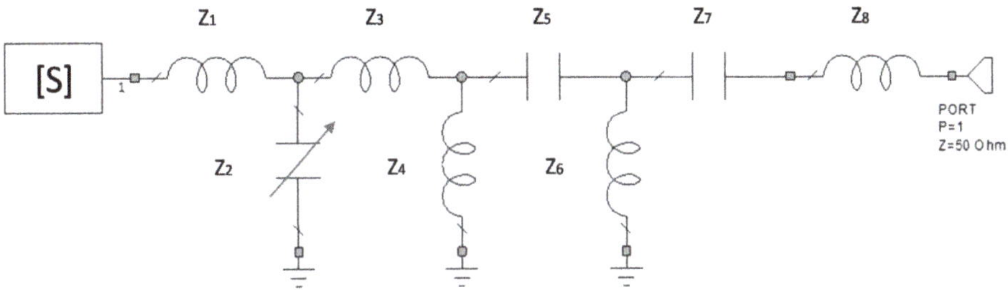

Figure 7. Reconfigurable matching network where Z_2 is the DTC. $Z_1 = 8$ nH, $Z_3 = 7.3$ nH, $Z_4 = 11$ nH, $Z_5 = 0.8$ pF, $Z_6 = 13$ nH, $Z_7 = 1.9$ pF, and $Z_8 = 2.5$ nH. All passive components were SMD (Surface-Mount Device) 0402 type, with a high Q and tight tolerance (from Murata).

The state of the DTC was software-controlled with a parallel interface connection at the end of the ground plane. Since the ground plane was a relevant element for the radiation process, to avoid an undesired effect of the interface connection, the following set-up was prepared. The DTC was DC-powered with a small battery embedded on the ground plane (Figure 6). Then, a particular state of the DTC was selected, having the interface connection attached. Once the DTC was operating at a given state, the interface connection was removed. Since the DTC received DC voltage from the battery, the state was maintained, and thus both S_{11} and total efficiency measurements could be carried out without the impact of the interface connection. The typical power consumption of the DTC was 180 µW, which represented about 2% of the power of a transceiver.

After optimizing the matching network, some relevant states were chosen to cover the range of 698–960 MHz up to 1710–2690 MHz (Figure 8).

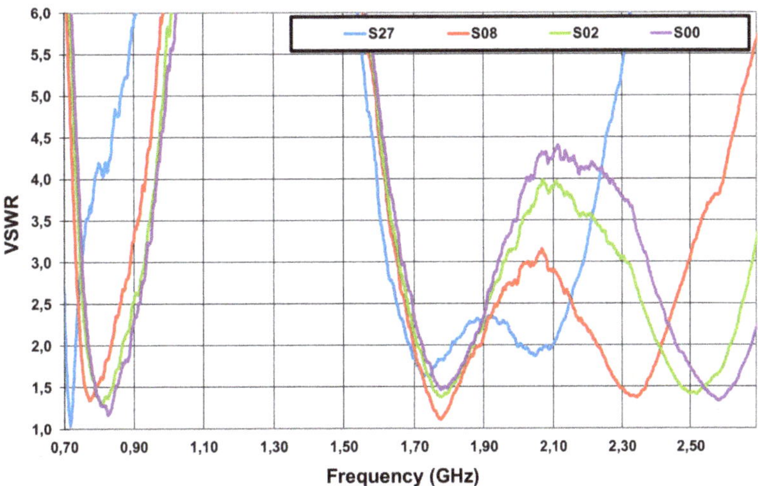

Figure 8. Measured VSWR for four states: S27, S08, S02, and S00.

It was shown from the VSWR measurements that a particular state (S27) could cover some low bands of 700 MHz, namely those from 698 MHz to 746 MHz. From 746 MHz up to 824 MHz, the S08 state could be selected. It is remarkable that four states were almost enough to obtain an acceptable VSWR for the whole frequency range (Figure 8).

To analyze the radiation performance, the total efficiency (η_t) was measured inside an anechoic chamber using 3D pattern integration (Satimo Stargate-32). The total efficiency considered both the radiation efficiency (η_r) and the mismatch losses as follows: $\eta_t = \eta_r \times (1 - |S_{11}|^2)$ (Figure 9). The gain results were also measured using the same set-up, resulting in an average gain of 1.5 dBi and 1.1 dBi at the low-frequency region and the high-frequency region, respectively. Radiation patterns are like those of a half-wavelength dipole, which are suitable for wireless environments since the wireless device may receive radio signals from any direction.

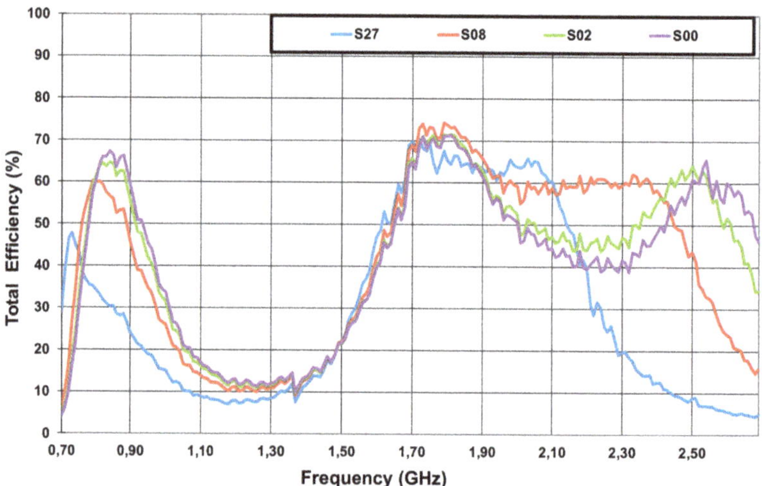

Figure 9. Measured total efficiency for the same four states shown in Figure 8.

In order to summarize the results, the average total efficiency values for several representative communication systems, namely LTE-12 (699–746 MHz), LTE-13, LTE-14 (746–798 MHz), LTE-5 (824–894 MHz), LTE-8 (880–960 MHz), LTE-3 (1710–1880 MHz), LTE-2 (1850–1990 MHz), LTE-1 (1920–2170 MHz), LTE-30 (2305–2360 MHz), and LTE-41 (2496–2690 MHz), are gathered in Table 1. The information in Table 1 helped to choose the best state for a specific LTE band, shown in a dark color. For example, regarding the low frequency region, if operation were needed at LTE-12, the S27 state should be selected, providing an average efficiency value of 41.5%. If LTE-13–14 were desired, then state S08 should be selected. For LTE-5 and 8, S00 could be selected. The same criteria applied for the high-frequency region. For example, to cover LTE-3, state S08 could be selected. S08 could be chosen for covering LTE-2, 1 and 30. Finally, S00 could be selected for covering LTE-41.

Due to the limitations of the anechoic chamber below 700 MHz, only measurements up to state S27 were carried out. However, tuning could go below 700 MHz, since more states were available.

To sum up, a small antenna booster of 86.4 mm^3 and a reconfigurable matching network allowed coverage in the frequency bands from 698 MHz to 960 MHz and from 1710 MHz to 2690 MHz. To achieve competitive performance, the reconfigurable matching network was designed using a MEMS tunable capacitor (DTC) combined with lumped inductors and capacitors. The measured average total efficiency was above 40% and 57% at the 698–960 MHz and 1710–2690 MHz frequency ranges, respectively, making the architecture attractive for multiband wireless devices.

Table 1. Measured total efficiencies for four states. The shadow indicates the best state for each frequency region.

	12	13, 14	5	8	3	2	1	30	41
State	699–746	746–798	824–894	880–960	1710–1880	1850–1990	1920–2170	2305–2360	2496–2690
S27	41.5	37.7	29.0	32.4	65.8	65.3	59.9	17.2	6.0
S08	19.0	53.6	54.0	41.5	71.5	63.8	59.1	60.4	26.2
S02	13.0	46.8	62.5	50.2	69.0	59.6	50.1	48.8	50.6
S00	11.5	44.1	65.4	53.7	68.4	58.5	47.7	42.5	57.2

4. RF Switches for Antenna Boosters

The DTC-based matching network for antenna boosters shown in the previous section is very attractive when only one component (a capacitor) of the matching network can tune the frequency bands of operation. However, in other situations, changing only the value of a capacitor is not enough to achieve satisfactory performance at the frequency bands of interest. This is the case of the following example illustrating a small platform (50 mm × 50 mm) to operate at several cellular frequency bands from 698 MHz to 960 MHz and from 1710 MHz to 2200 MHz, as well as at 1575 MHz for geolocalization (Figure 10).

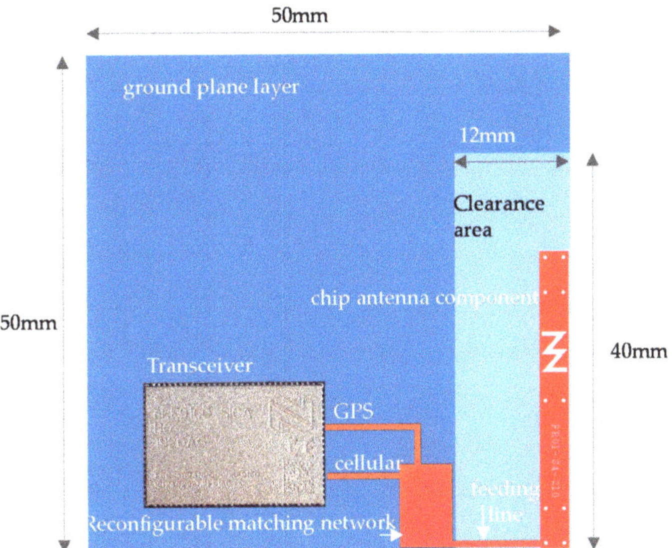

Figure 10. An antenna booster component (in red) on a PCB (50 mm × 50 mm) connected to a reconfigurable matching network with single-pole n-throw (SPNT) switches connected to a transceiver for cellular and GPS communication.

In the present case, to gain more freedom, the multiband operation was addressed with simple matching networks comprising one or two lumped components designed for matching a particular frequency band. In this regard, a set of matching networks were connected using two single-pole N-throw (SPNT) switches as shown in the coming sections.

4.1. SPNT Switch

SPNT switches permit connecting a common radiofrequency input (pole) to several output ports (throws) (Figure 11). Besides the number of throws, insertion loss and isolation form an important figure of merit, since it impacts both the transmitted and received signal strength [37]. For the present case, the insertion loss of the SP8T switch is 0.5 dB and 0.7 dB at 698 MHz and 2170 MHz, respectively, with a power handling of 32 dBm. Isolation is typically better than 40 dB at the whole frequency range of operation.

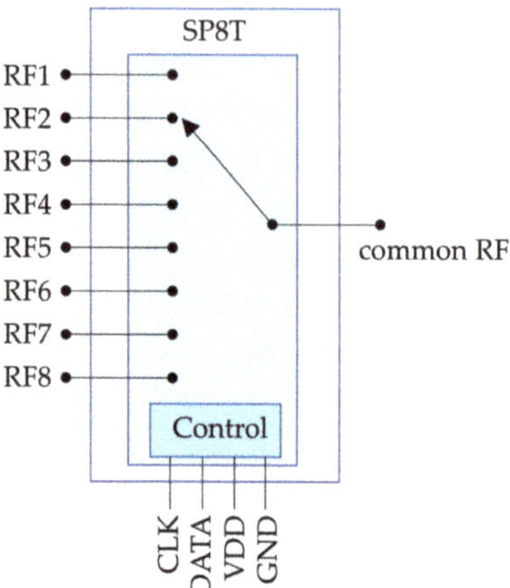

Figure 11. An SP8T switch includes a common RF port, eight throws, and a control interface to set a connection between an RF_i and the common RF.

For an SPNT switch, N can adopt values from 2 up to 12, where each state can be controlled either with a general-purpose input/output (GPIO) or an RFFE interface. The GPIO interface comprises control lines to set the state of the switch, based on a simple DC voltage-driven scheme. For example, for an SP8T, three control lines are needed to control eight states: RF1 using 001, RF2 using 010, and so on. RFFE simplifies the number of control lines (1 clock and 1 data) and employs a single-ended CMOS input/output port, minimizing power consumption [38].

In the following section, a pair of SP8T switches having 2 mm $\times$ 2 mm size are included to provide operation at both cellular and GPS bands using an antenna booster of only 90 mm^3.

4.2. Multiband Operation with an SP8T Switch

A multiband reconfigurable architecture comprising an antenna booster with a bunch of matching networks is proposed for operation at cellular bands, as well as GPS (Figure 12) [39]. A first simulation analysis was carried out, simulating a PCB as shown in Figure 10, including an antenna booster of 30 mm $\times$ 3 mm $\times$ 1 mm (h). Once the S-parameters at the antenna port were obtained, the matching network architecture, including the passive matching networks and the RF switches, was designed. Once the design was feasible from the simulation domain, the solution was implemented into a real cellular IoT device (Figure 13).

The cellular frequency band of operation was automatically controlled by a transceiver (by Nordic Semiconductor) through a GPIO interface, although an RFFE interface could be used as well. Such a GPIO interface controls the SP8T switches with three control lines (CLT1–3). In this case, six matching networks were designed with circuit simulator software (Cadence AWR Microwave Office) including the measured S-parameters of the switches as well as the S-parameters of the SMD components of the matching network. Those matching networks were used to match different cellular bands as well as GPS. For each frequency band, a simple L-type matching network using SMD components was used (Figure 14).

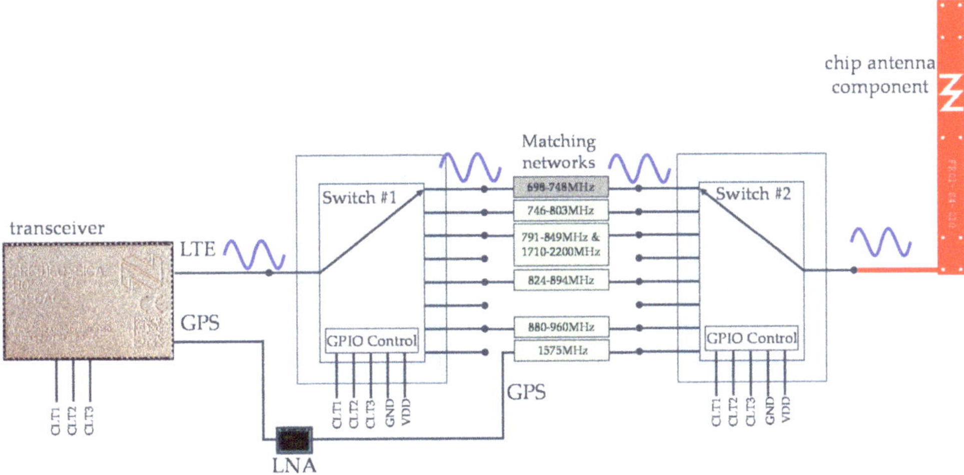

Figure 12. Reconfigurable matching network system for multiband operation. A transceiver (cellular and GPS) controls the SP8T switches through lines CLT1–3, matching networks comprising lumped SMD components (capacitors and inductors). A low noise amplifier (LNA) amplifies the GPS signal before entering the transceiver.

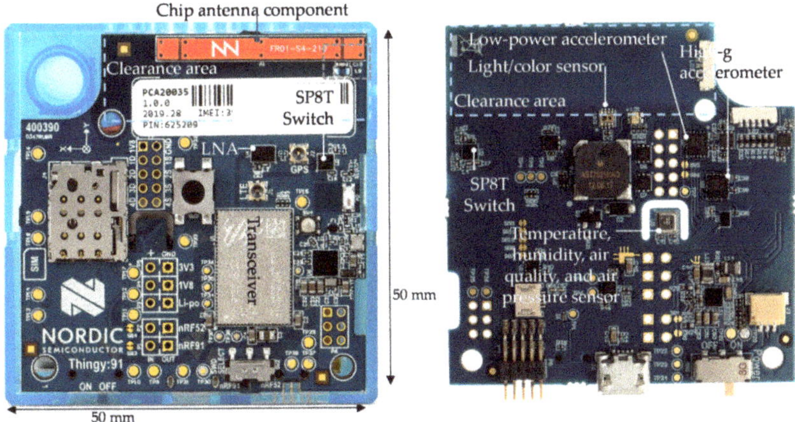

Figure 13. A 50 mm × 50 mm IoT open platform device (Thingy:91 by Nordic Semiconductor), including the reconfigurable antenna system of Figure 12. Besides cellular and GPS operation thanks to the reconfigurable antenna system (by Ignion), many sensors for IoT applications are included, such as temperature, humidity, air quality, air pressure, color, and light data [39].

To keep the solution as compact as possible at the cellular bands, SMD 0201 (0.5 mm × 0.25 mm) lumped capacitors and inductors were used for the matching networks, selecting the highest quality factor (Q) possible. For the 1575 MHz matching network, SMD 0402-lumped components were used to achieve a higher Q (Figure 14).

Each matching network provided operation at a single band (RF1, RF3, or RF4) or dual band (RF5, RF7, or RF8) with an S_{11} value that was optimized within the band (Figure 15).

STATE	Frequency band	Matching Network		
RF5	698-748MHz & 1710-2200MHz	Component	Value	Q_{min}
		Z_1	5.1nH	40 at 698MHz
		Z2 (C10)	5.5pF	76 at 2200MHz
RF7	746-803MHz& 1710-2200MHz	Component	Value	Q_{min}
		Z_1	4.7nH	40 at 746MHz
		Z_2	0Ω	-
RF8	791-849MHz& 1710-2200MHz	Component	Value	Q_{min}
		Z_1	9.2pF	166 at 849MHz
		Z_2	6.2nH	40 at 791MHz
RF3	824-894MHz	Component	Value	Q_{min}
		Z_1	1.5pF	461 at 894MHz
RF1	880-960MHz	Component	Value	Q_{min}
		Z_1	2.5pF	402 at 960MHz
		Z_2	Open circuit	-
RF4	GPS (1575MHz)	Component	Value	Q_{min}
		Z_1	2.2nH	103 at 1575MHz
		Z_2	2.5pF	242 at 1575MHz
RF2&6	available for other bands	empty		

Figure 14. Matching networks for cellular and GPS, used in the electric scheme shown in Figure 12 where the box with [S] includes the S_{11} of switch #2 and the antenna booster.

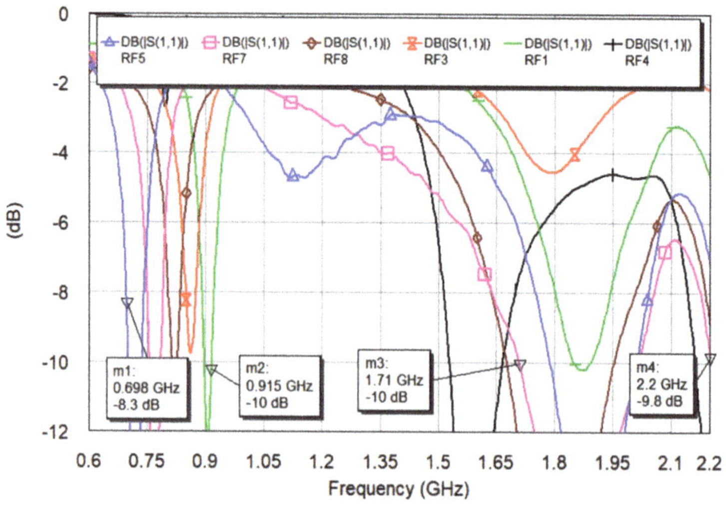

Figure 15. S_{11} for several states covering cellular and GPS bands. Each state corresponds to one matching network as shown in Figure 14.

For each state, the total efficiency was measured in an anechoic chamber using 3D pattern integration. As stated before, total efficiency considered mismatching as well as losses of the system, such as those of the antenna system, and the nearby components (e.g., the lithium polymer battery under the PCB) as well as the top and bottom plastic covers (Table 2).

Table 2. Measured total efficiency for the IoT device shown in Figure 13, including all components and top and bottom plastic casings.

Band (MHz)	699–748	746–803	791–849	824–894	880–960	1575	1710–2170
Efficiency (%)	10.0	12.6	15.7	18.5	11.2	39.8	47.4

5. Discussion

This section discusses several reconfigurable antenna solutions that have appeared in industry, as well as the scientific literature [40–54]. The selection criteria included those reconfigurable antenna solutions for multiband behavior at two frequency regions: a first one from 698 MHz to 960 MHz and a second one from 1.71 GHz to at least 2.17 GHz. The first frequency region was by far the most challenging for designers, since the antennas were electrically smaller in this first frequency region than in the second. Consequently, it was more difficult to squeeze bandwidth from the first frequency region.

The antennas were classified by length and width. In those antennas having irregular shapes, the length and width were selected as the minimum rectangle comprising the antenna. Once all antennas were put in a single map, a length and width average could be calculated. That single average antenna defined four quadrants: the best, the worst, and two intermediate quadrants, one having a length below the average but widths above and the other way around for the other intermediate quadrant (Figure 16).

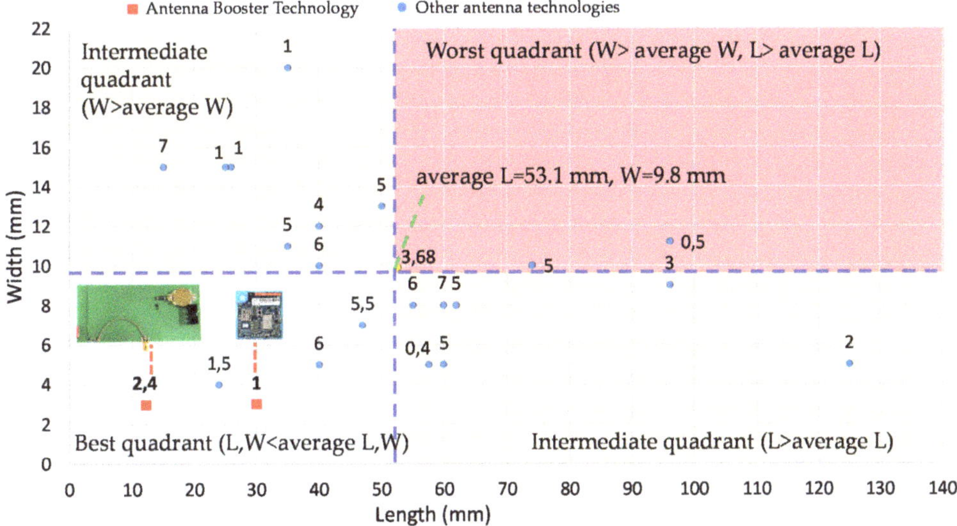

Figure 16. Classification of several reconfigurable multiband antennas.

Regarding the active component for the matching network, not only were DTCs and SPNT switches used, but some solutions included PIN diodes to connect or disconnect one or more conductive parts of the antenna to obtain a multiband performance.

For comparison purposes, the two-antenna booster's reconfiguration-based solutions presented in this paper were included in the plot. The one using an antenna booster of 12 mm × 3 mm × 2.4 mm, providing operation from 698 MHz to 960 MHz and from 1710 MHz to 2690 MHz, and a second one featuring an antenna booster of 30 mm × 3 mm × 1 mm for operation at 698–960 MHz and 1710–2170 MHz, as well as in GPS using the same reconfigurable matching circuit. Besides the lengths and widths, the antenna height was included in the plot with a number indicating heights in millimeters.

As is clearly shown in Figure 16, the antenna booster's reconfiguration-based solutions were in the best quadrant, and furthermore, they were in the best location within that area; the one with a 2.4 mm height in particular was the best among all samples, featuring a small volume of only 86.4 mm^3. In second place was the 30 mm $\times$ 3 mm $\times$ 1 mm antenna booster featuring a volume of 90 mm^3. This solution had a smaller volume than the one having a 1.5 mm height, having a volume of 144 mm^3. As a final remark, antenna booster technology features the smallest footprint and volume while providing performance across many frequencies within a 0.7–3 GHz range.

6. Conclusions

Two multiband reconfigurable antenna architectures for wireless devices using antenna boosters have been presented, having a small antenna booster size of less than 90 mm^3 for operation at 698–960 MHz as well as 1710–2690 MHz, being the smallest volume among twenty antenna samples from the prior-art. Such multiband reconfigurability was achieved by combining digitally tunable capacitors and RF switches with antenna boosters. On one hand, the DTC provides reconfigurability thanks to a tunable capacitor that, when combined with a multiband matching network and an antenna booster, allows reconfiguration of the frequency bands. In other situations where changing a capacitor is not sufficient for providing enough freedom to tune the frequency bands, an architecture with RF switches, such as the one presented here, can be adopted. Both DTCs and RF switches use an interface than can be easily connected with most transceivers. Two solutions have been implemented: one with a DTC and an antenna booster of 86.4 mm^3 in a reference board and another with two RF switches and an antenna booster of 90 mm^3 in a cellular IoT device, showing competitive efficiency values for the wireless industry. These antenna booster architectures can be easily adopted in wireless devices because the antenna booster does not need geometry customization, since the multiband operation is adjusted by the design of the matching network, being a fast and easy design process. Finally, an antenna booster can simply be embedded due to its small size and off-the-shelf nature, resulting in attractive technology for the everything-connected world.

Author Contributions: J.A. and A.A. conceived the paper structure and, together with J.L.L. and O.M., led the development of the antenna booster part. J.T., E.R., and R.B. led the IoT device (Thingy:91) development, contributing to the RF switch section. R.G. contributed to the DTC section. All authors have read and agreed to the published version of the manuscript.

Funding: Part of this research was financed by the Spanish Ministry of Industry, Energy, and Tourism, belonging to the National Plan of Scientific Research, Development, and Technology Innovation 2015–2017 (Project Ref: TSI-100103-2015-39). It was also financed by the CDTI (Centro Desarrollo Tecnológico Industrial) of the Spanish Ministry of Science and Innovation in 2020–2021 (Project Ref: TIC-20200069).

Conflicts of Interest: The authors declare no conflict of interest.

References

1. Anguera, J.; Andújar, A.; Huynh, M.C.; Orlenius, C.; Picher, C.; Puente, C. Advances in Antenna Technology for Wireless Handheld Devices. *Int. J. Antennas Propag.* **2013**, *2013*, 838364. [CrossRef]
2. Zheng, M.; Wang, H.; Hao, Y. Internal Hexa-Band Folded Monopole/Dipole/Loop Antenna With Four Resonances for Mobile Device. *IEEE Trans. Antennas Propag.* **2012**, *60*, 2880–2885. [CrossRef]
3. Kim, B.-N.; Park, S.-O.; Lee, J.-H.; Oh, J.-K.; Lee, K.-J.; Koo, G.-Y. Hepta-band planar inverted-F antenna with novel feed structure for wireless terminals. In Proceedings of the 2007 IEEE Antennas and Propagation Society International Symposium, Honolulu, HI, USA, 9–15 June 2007; pp. 1257–1260.
4. Ban, Y.-L.; Qiang, Y.-F.; Chen, Z.; Kang, K.; Li, J.L.-W. Low-Profile Narrow-Frame Antenna for Seven-Band WWAN/LTE Smartphone Applications. *IEEE Antennas Wirel. Propag. Lett.* **2014**, *13*, 463–466. [CrossRef]
5. Chen, Z.X.; Ban, Y.L.; Chen, Z.; Kang, K.; Li, J.L.W. Two-strip narrow-frame monopole antenna with a capacitor loaded for hepta-band smartphone applications. *Prog. Electromagn. Res.* **2014**, *145*, 31–38. [CrossRef]
6. Ban, Y.-L.; Liu, C.-L.; Li, J.L.-W.; Li, R. Small-Size Wideband Monopole With Distributed Inductive Strip for Seven-Band WWAN/LTE Mobile Phone. *IEEE Antennas Wirel. Propag. Lett.* **2013**, *12*, 7–10. [CrossRef]

7.   Deng, C.; Li, Y.; Zhang, Z.; Feng, Z. A Novel Low-Profile Hepta-Band Handset Antenna Using Modes Controlling Method. *IEEE Trans. Antennas Propag.* **2014**, *63*, 799–804. [CrossRef]
8.   Wang, Y.; Du, Z. Wideband Monopole Antenna With Less Nonground Portion For Octa-Band WWAN/LTE Mobile Phones. *IEEE Trans. Antennas Propag.* **2016**, *64*, 383–388. [CrossRef]
9.   Huang, D.; Du, Z.; Wang, Y. An Octa-band Monopole Antenna With a Small Nonground Portion Height for LTE/WLAN Mobile Phones. *IEEE Trans. Antennas Propag.* **2016**, *65*, 878–882. [CrossRef]
10.   Wu, D.; Cheung, S.W.; Yuk, T.I. A Compact and Low-Profile Loop Antenna With Multiband Operation for Ultra-Thin Smartphones. *IEEE Trans. Antennas Propag.* **2015**, *63*, 2745–2750. [CrossRef]
11.   Xu, Z.-Q.; Zhou, Q.-Q.; Ban, Y.-L.; Ang, S.S. Hepta-Band Coupled-Fed Loop Antenna For LTE/WWAN Unbroken Metal-Rimmed Smartphone Applications. *IEEE Antennas Wirel. Propag. Lett.* **2017**, *17*, 311–314. [CrossRef]
12.   Anguera, J.; Andújar, A.; Puente, C. Antenna-Less Wireless: A Marriage Between Antenna and Microwave Engineering. *Microw. J.* **2017**, *60*, 22–36.
13.   Anguera, J.; Andújar, A.; Puente, C.; Mumbrú, J. Antennaless Wireless Device Capable of Operation in Multiple Frequency Regions. U.S. Patent 8,736,497, 4 August 2008.
14.   Andújar, A.; Anguera, J.; Puente, C. Ground Plane Boosters as a Compact Antenna Technology for Wireless Handheld Devices. *IEEE Trans. Antennas Propag.* **2011**, *59*, 1668–1677. [CrossRef]
15.   Anguera, J.; Andújar, A.; Puente, C. Wireless Handheld Devices, Radiation Systems and Manufacturing Methods. Patent CN 104508905 (B), 19 July 2013.
16.   Anguera, J.; Picher, C.; Bujalance, A.; Andújar, A. Ground plane booster antenna technology for smartphones and tablets. *Microw. Opt. Technol. Lett.* **2016**, *58*, 1289–1294. [CrossRef]
17.   Anguera, J.; Andújar, A.; Mestre, G.; Rahola, J.; Juntunen, J. Design of Multiband Antenna Systems for Wireless Devices Using Antenna Boosters. *IEEE Microw. Mag.* **2019**, *20*, 102–114. [CrossRef]
18.   De Mingo, J.; Valdovinos, A.; Crespo, A.; Navarro, D.; Garcia, P. An RF Electronically Controlled Impedance Tuning Network Design and Its Application to an Antenna Input Impedance Automatic Matching System. *IEEE Trans. Microw. Theory Tech.* **2004**, *52*, 489–497. [CrossRef]
19.   Yang, S.; Zhang, C.; Pan, H.K.; Fathy, A.E.; Nair, V.K. Frequency-Reconfigurable Antennas for Multiradio Wireless Platforms. *IEEE Microw. Mag.* **2009**, *10*, 66–83. [CrossRef]
20.   Chi, Y.-W.; Wong, K.-L. Quarter-Wavelength Printed Loop Antenna With an Internal Printed Matching Circuit for GSM/DCS/PCS/UMTS Operation in the Mobile Phone. *IEEE Trans. Antennas Propag.* **2009**, *57*, 2541–2547. [CrossRef]
21.   Iyer, V.; Makarov, S.N.; Harty, D.D.; Nekoogar, F.; Ludwig, R. A Lumped Circuit for Wideband Impedance Matching of a Non-Resonant, Short Dipole or Monopole Antenna. *IEEE Trans. Antennas Propag.* **2010**, *58*, 18–26. [CrossRef]
22.   Li, Y.; Zhang, Z.; Zheng, J.; Feng, Z.; Iskander, M.F. A Compact Hepta-Band Loop-Inverted F Reconfigurable Antenna for Mobile Phone. *IEEE Trans. Antennas Propag.* **2012**, *60*, 389–392. [CrossRef]
23.   Liu, Y.; Zhou, Y.-M.; Liu, G.-F.; Gong, S.-X. Heptaband Inverted-F Antenna for Metal-Rimmed Mobile Phone Applications. *IEEE Antennas Wirel. Propag. Lett.* **2015**, *15*, 996–999. [CrossRef]
24.   Xu, H.; Wang, H.; Gao, S.; Zhou, H.; Huang, Y.; Xu, Q.; Cheng, Y. A Compact and Low-Profile Loop Antenna With Six Resonant Modes for LTE Smartphone. *IEEE Trans. Antennas Propag.* **2016**, *64*, 3743–3751. [CrossRef]
25.   Asadallah, F.A.; Costantine, J.; Tawk, Y. A Multiband Compact Reconfigurable PIFA Based on Nested Slots. *IEEE Antennas Wirel. Propag. Lett.* **2018**, *17*, 331–334. [CrossRef]
26.   Chen, Y.; Martens, R.; Valkonen, R.; Manteuffel, D. Evaluation of adaptive impedance tuning for reducing the form factor of handset antenas. *IEEE Trans. Antennas Propag.* **2015**, *63*, 703–710. [CrossRef]
27.   Hoarau, C.; Corrao, N.; Arnould, J.D.; Ferrari, P.; Xavier, P. Complete design and measurement methodology for a tunable RF impedance-matching network. *IEEE Trans. Microw. Theory Tech.* **2008**, *56*, 2620–2627. [CrossRef]
28.   Chen, Y.; Manteuffel, D. A Tunable Decoupling and Matching Concept for Compact Mobile Terminal Antennas. *IEEE Trans. Antennas Propag.* **2017**, *65*, 1570–1578. [CrossRef]
29.   Ida, I.; Takada, J.-I.; Toda, T.; Oishi, Y. An adaptive impedance matching system and its application to mobile antennas. In Proceedings of the 2004 IEEE Region 10 Conference TENCON 2004, Chiang Mai, Thailand, 21–24 November 2004; pp. 543–546.
30.   Bahramzy, P.; Olesen, P.; Madsen, P.; Bojer, J.; del Barrio, S.C.; Tatomirescu, A.; Bundgaard, P.; Iii, A.S.M.; Pedersen, G.F. A Tunable RF Front-End With Narrowband Antennas for Mobile Devices. *IEEE Trans. Microw. Theory Tech.* **2015**, *63*, 3300–3310. [CrossRef]
31.   Anguera, J.; Toporcer, N.; Andújar, A. Slim Bar Boosters for Electronic Devices. U.S. Patent 9,960,478, 24 July 2014.
32.   Andújar, A.; Anguera, J.; Mateos, R. Multiband non-resonant antenna system with reduced ground clearance. In Proceedings of the 11th European Conference on Antennas and Propagation (EUCAP), Paris, France, 19–24 March 2017; pp. 3086–3089.
33.   Fano, R.M. Theoretical Limitations on the Broad-Band Matching of Arbitrary Impedances. *J. Frankl. Inst.* **1950**, *249*, 57–83. [CrossRef]
34.   Anguera, J.; Andújar, A.; Leiva, J.; Schepens, C.; Gaddi, R.; Kahng, S. Multiband antenna operation with a non-resonant element using a reconfigurable matching network. In Proceedings of the 12th European Conference on Antennas and Propagation (EuCAP 2018), London, UK, 9–13 April 2018.
35.   Tiggelman, M.P.J.; Reimann, K.; van Rijs, F.; Schmitz, J.; Hueting, R.J.E. On the Trade-Off Between Quality Factor and Tuning Ratio in Tunable High-Frequency Capacitors. *IEEE Trans. Electron. Devices* **2009**, *56*, 2128–2136. [CrossRef]

36. Anguera, J.; Andújar, A. Wireless Device. U.S. Patent 10,122,403, 12 January 2016.
37. Wang, D.; Wolf, R.; Joseph, A.; Botula, A.; Rabbeni, P.; Boenke, M.; Harame, D.; Dunn, J. High performance SOI RF switches for wireless applications. In Proceedings of the 10th IEEE International Conference on Solid-State and Integrated Circuit Technology, Shanghai, China, 1–4 November 2010; pp. 611–614.
38. MIPI Alliance. *MIPI White Paper: Enabling the IoT Opportunity*; MIPI Alliance: Piscataway, NJ, USA, 2020.
39. Anguera, J.; Andújar, A.; Leiva, J.L.; Cobo, Y.; Tønnessen, J.; Rindalsholt, E.; Brandsegg, R. *Virtual Antenna™ Provides Mobile and GPS Connection in the Thingy:91 Cellular IoT Module*; Fractus Antennas: Barcelona, Spain, 2020.
40. Trinh, L.H.; Ferrero, F.; Lizzi, L.; Staraj, R.; Ribero, J.-M. Reconfigurable Antenna for Future Spectrum Reallocations in 5G Communications. *IEEE Antennas Wirel. Propag. Lett.* **2016**, *15*, 1297–1300. [CrossRef]
41. Del Barrio, S.C.; Foroozanfard, E.; Morris, A.; Pedersen, G.F. Tunable Handset Antenna: Enhancing Efficiency on TV White Spaces. *IEEE Trans. Antennas Propag.* **2017**, *65*, 2106–2111. [CrossRef]
42. Mun, B.; Jung, C.; Park, M.-J.; Lee, B. A Compact Frequency-Reconfigurable Multiband LTE MIMO Antenna for Laptop Applications. *IEEE Antennas Wirel. Propag. Lett.* **2014**, *13*, 1389–1392. [CrossRef]
43. Ban, Y.-L.; Sun, S.-C.; Li, P.-P.; Li, J.L.-W.; Kang, K. Compact Eight-Band Frequency Reconfigurable Antenna for LTE/WWAN Tablet Computer Applications. *IEEE Trans. Antennas Propag.* **2014**, *62*, 471–475. [CrossRef]
44. Ilvonen, J.; Valkonen, R.; Holopainen, J.; Viikari, V. Multiband Frequency Reconfigurable 4G Handset Antenna with MIMO Capability. *Prog. Electromagn. Res.* **2014**, *148*, 233–243. [CrossRef]
45. Peng, C.-M.; Chen, I.-F.; Liu, C.-H. Multiband Printed Asymmetric Dipole Antenna for LTE/WLAN Applications. *Int. J. Antennas Propag.* **2013**, *2013*, 1–6. [CrossRef]
46. Zhang, C.; Yang, S.; El-Ghazaly, S.; Fathy, A.; Nair, V. A Low-Profile Branched Monopole Laptop Reconfigurable Multiband Antenna for Wireless Applications. *IEEE Antennas Wirel. Propag. Lett.* **2008**, *8*, 216–219. [CrossRef]
47. Hu, C.-L.; Huang, D.-L.; Kuo, H.-L.; Yang, C.-F.; Liao, C.-L.; Lin, S.-T. Compact Multibranch Inverted-F Antenna to be Embedded in a Laptop Computer for LTE/WWAN/IMT-E Applications. *IEEE Antennas Wirel. Propag. Lett.* **2010**, *9*, 838–841. [CrossRef]
48. Li, H.; Xiong, J.; Yu, Y.; He, S. A Simple Compact Reconfigurable Slot Antenna With a Very Wide Tuning Range. *IEEE Trans. Antennas Propag.* **2010**, *58*, 3725–3728. [CrossRef]
49. Li, Y.; Zhang, Z.; Zheng, J.; Feng, Z. Compact Heptaband Reconfigurable Loop Antenna for Mobile Handset. *IEEE Antennas Wirel. Propag. Lett.* **2011**, *10*, 1162–1165. [CrossRef]
50. Ren, Y.-J. Ceramic Based Small LTE MIMO Handset Antenna. *IEEE Trans. Antennas Propag.* **2012**, *61*, 934–938. [CrossRef]
51. Dioum, I.; Diallo, A.; Farssi, S.M.; Luxey, C. A Novel Compact Dual-Band LTE Antenna-System for MIMO Operation. *IEEE Trans. Antennas Propag.* **2014**, *62*, 2291–2296. [CrossRef]
52. Singh, A.; Shamblin, J.; Jones, R.; Rowson, S.; Pajona, O.; Sron, S.T.; Floc'h, J.M.; Drissi, M.; Singh, A.; Jones, R. Compact active antenna for mobile devices supporting 4G LTE. In Proceedings of the 2014 Loughborough Antennas and Propagation Conference (LAPC), Loughborough, UK, 10–11 November 2014; pp. 525–529.
53. Choi, M.; Wi, H.; Mun, B.; Yoon, Y.; Lee, H.; Lee, B. A Compact Frequency Reconfigurable Antenna for LTE Mobile Handset Applications. *Int. J. Antennas Propag.* **2015**, *2015*, 1–10. [CrossRef]
54. Houret, T.; Lizzi, L.; Ferrero, F.; Danchesi, C.; Boudaud, S. DTC-Enabled Frequency-Tunable Inverted-F Antenna for IoT Applications. *IEEE Antennas Wirel. Propag. Lett.* **2019**, *19*, 307–311. [CrossRef]

Article

Characterization of Novel Structures Consisting of Micron-Sized Conductive Particles That Respond to Static Magnetic Field Lines for 4G/5G (Sub-6 GHz) Reconfigurable Antennas

Adnan Iftikhar [1,†] , Jacob M. Parrow [2,†], Sajid M. Asif [3,†], Adnan Fida [1,†], Jeffery Allen [4,†], Monica Allen [4,†], Benjamin D. Braaten [2,†] and Dimitris E. Anagnostou [5,*,†]

1   Department Electrical and Computer Engineering, COMSATS University Islamabad,
    Islamabad 45550, Pakistan; adnaniftikhar@comsats.edu.pk (A.I.); adnan_fida@comsats.edu.pk (A.F.)
2   Department of Electrical and Computer Engineering, North Dakota State University, Fargo, ND 58102, USA;
    jacob.parrow@ngc.com (J.M.P.); benjamin.braaten@ndsu.edu (B.D.B.)
3   Electronic and Electrical Engineering, The University of Sheffield, Sheffield S1 4ET, UK; sajid.asif@gmail.com
4   Air Force Research Laboratory, Munitions Directorate, Eglin AFB, FL 32542, USA;
    jeffery.allen.12@us.af.mil (J.A.); monica.allen.3@us.af.mil (M.A.)
5   Department of Electrical Engineering, Heriot-Watt University, Edinburgh EH14 4AS, UK
*   Correspondence: d.anagnostou@hw.ac.uk
†   These authors contributed equally to this work.

Received: 16 April 2020; Accepted: 20 May 2020; Published: 29 May 2020

Abstract: Controlling Radio Frequency (RF) signals through switching technology is of interest to designers of modern wireless platforms such as Advanced Wireless services (AWS) from 2.18 GHz–2.2 GHz, mid-bands of sub-6 GHz 5G (2.5 GHz and 3.5 GHz), and 4G bands around 600 MHz/700 MHz, 1.7 GHz/2.1 GHz/2.3 GHz/2.5 GHz. This is because certain layout efficiencies can be achieved if suitable components are chosen to control these signals. The objective of this paper is to present a new model of an RF switch denoted as a Magnetostatic Responsive Structure (MRS) for achieving reconfigurable operation in 4G/5G antennas. In particular, the ABCD matrices of the MRS are derived from the S-parameter values and shown to be a good model from 100 kHz to 3.5 GHz. Furthermore, an overall agreement between simulations, analytical results, and circuit model values are shown.

Keywords: solid state switches; RF switching; magnetic particles; magnetostatic responsive structures (MRSs)

---

1. Introduction

The use of RF switches is becoming crucial in RF designs as the mobile industry is moving toward faster data-rates and having flexibility of more wireless services with lower power consumption. There are three existing state-of-the-art solid-state RF switching technologies: PIN diodes, micro-electro-mechanical system (MEMS), and solid-state field-effect transistors (FETs) [1,2]. PIN diodes are suitable for high power applications and require a large biasing current with slow switching speed. In addition, MEMS RF switches [3] have low insertion loss and high isolation, but their special packaging, large actuation voltage requirement, and reliability issues are major concerns and restricts their use in RF switching applications [1]. Solid-state FETs are reliable and require low power consumption, but at higher frequencies there exists serious trade-offs between isolation and insertion loss characteristics of FETs.

Furthermore, much of the existing RF switching technology has miniature sizes but they require direct current (DC) biasing circuitry for their operation and isolation between RF and DC currents [1]. Therefore, the incorporation of a DC biasing network increases the overall size of the RF switch and restricts their embedment on the RF instrumentation substrate. On the other hand, RF reed switches [4] do not need any biasing circuitry; however, their size is relatively bigger and requires a large magnetic force for their operation [4–6]. On the contrary, Magnetostatic Responsive Structures (MRSs) require a small magnetic force for their operation and are compact in size, which makes them a good candidate for embedment on the RF instrumentation substrate.

The objective of this paper is to present detailed insertion loss and isolation characteristics of the Magnetostatic Responsive Structures (MRSs), shown in Figure 1, using ABCD parameters from the measured S-parameters and equivalent lumped element model. The magnetic particles were manufactured by Potters Industries LLC [7] and a microscopic view is shown in Figure 1a. The substrate used for the demonstration of the MRSs was Rogers TMM4 [8]. The behavior of the micron-size magnetic particles in the presence of a static magnetic field is shown in Figure 1b, whereas the parameters and architecture of the MRSs are shown in Figure 1c. Two different sizes of MRSs, as shown in Figure 1d,e, with a cavity diameter of 0.9 mm were filled with the calculated amount of the micron-size magnetic particles having an average diameter of 20 μm–40 μm. The comparison of the insertion loss and isolation characteristics showed that both MRSs have RF switching capabilities and the smaller the MRSs provided functionality from 100 kHz to 3.5 GHz, when a static magnetic field was applied and removed, respectively.

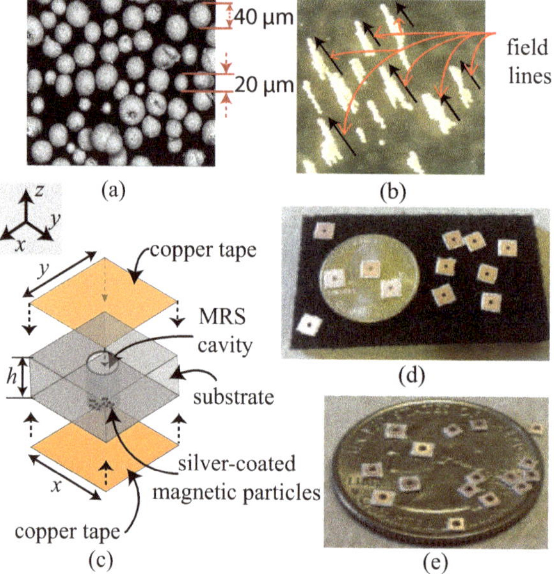

Figure 1. (**a**) microscopic view of silver coated magnetostatic responsive particles, (**b**) vertical alignment of magnetic particles in the presence of static magnetic field, (**c**) exploded view of novel Magnetostatic Responsive Structure (MRS), (**d**) manufactured MRSs with $h = 0.508$ mm , $x = y = 3$ mm, cavity diameter $d = 0.9$ mm, and (**e**) Manufactured MRS with cavity height $h = 0.508$ mm, $x = y = 1.5$ mm, and $d = 0.9$ mm.

The novel MRSs have recently been used to frequency reconfigure a printed dipole antenna [9] and also to achieve a reconfigurable band-stop filter [10]. In addition, the comparison between the MRSs and PIN diodes used for the frequency reconfiguration of a microstrip patch antenna in [11] revealed that avoidance of DC biasing circuitry for the MRSs increased the overall antenna performance. Moreover, the bandwidth exploration, signal integrity, benefits of remote biasing, and scalability issues of MRSs have also been reported in [12]. The applications of the Magnetostatic Responsive Structures (MRSs) [9–11] and signal integrity aspects [12] demonstrated that MRS is a potential alternative switching technology. However, a significant addition to RF switching technology and replacement of the conventional RF switches (PIN diodes, Reed switches, etc.) at low frequencies from 100 kHz to 3.5 GHz and even for higher frequencies could only be possible after complete realization and characterization of MRSs.

2. Development of Magnetostatic Responsive Structures (MRS)

The manufacturing of the MRS was achieved in-house using the LPKF ProtoMat S63 PCB [13] milling machine. In particular, milling practices were used to make the cavities in a substrate and for accurate cutting of the MRSs. Initially, a 0.9 mm cavity diameter d on a 3.0 mm $\times$ 3.0 mm substrate having a thickness of 0.508 mm h was accurately manufactured and is shown in Figure 1d. The substrate material used was a Rogers TMM4 with 1 oz. copper cladding on both the top and bottom. This practice validated the feasibility of milling practices and even smaller structures were manufactured using this milling machine process. Figure 1e shows the manufactured MRSs cavities with a diameter of 0.9 mm on a Rogers TMM4 substrate having a size of 1.5 mm $\times$ 1.5 mm $\times$ 0.508 mm.

After the manufacturing of the MRSs, the cavities of the MRSs were then filled with the calculated amount of micron-sized silver coated magnetic particles. The amount of the micron-sized particles was estimated using the Kepler's problem [14] for packing of the spherical balls in a cylindrical volume. The derived formula is given as:

$$N_t = \frac{\pi r^2 h}{4\sqrt{2}R^3},\tag{1}$$

where r is the radius of cylindrical drilled cavity, h is the height of the substrate, and R is the average radius of the silver coated magnetic particles.

Equation (1) was first used to estimate the amount of the silver coated magnetic particles in the cavity having a size smaller than the manufactured MRSs shown in Figure 1d,e. The smaller manufactured cavities were called a measuring cup and have dimensions of 1 mm $\times$ 1 mm $\times$ 0.508 mm and d = 0.4 mm size. The manufactured measuring cup is shown in Figure 2. This measuring cup was first fully filled with the magnetic particles and the amount of magnetic particles was estimated using Equation (1). A microscopic view of the measuring cup filled with micron-sized silver coated magnetic particles is shown in Figure 3. Then, the particles in the measuring cup were transferred to the MRSs cavities by first transferring to a sheet of paper and then migrated to cavities using a static magnet under the manufactured MRSs.

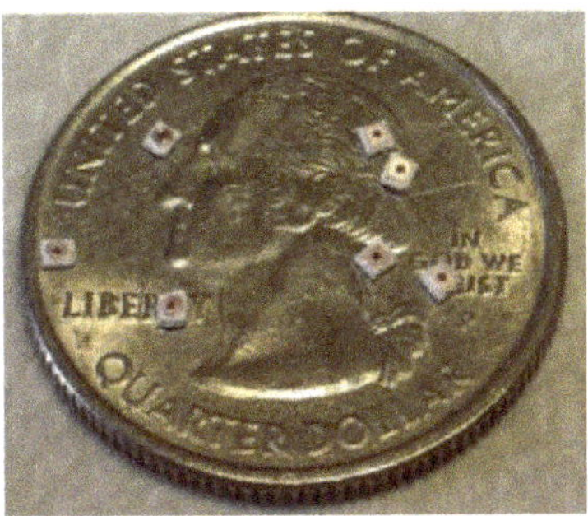

Figure 2. Manufactured measuring cups for the filling of magnetic particles in the MRS cavities.

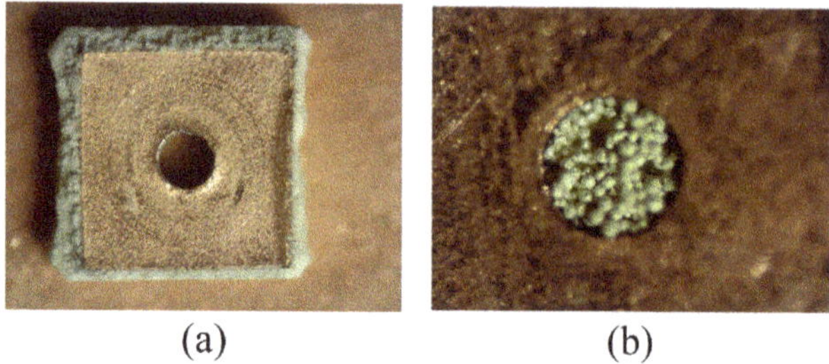

Figure 3. (**a**) a microscopic view of manufactured measuring cup with a cavity diameter of 0.4 mm on TMM4 substrate having dimensions 1 mm × 1.5 mm × 0.508 mm and (**b**) manufactured measured cup completely filled with the micron-sized magnetic particles.

3. Characterization of Magnetostatic Responsive Structures (MRSs) Using ABCD Parameters

Next, the insertion loss and isolation of the MRSs were extracted using the ABCD parameters by designing and prototyping a discontinuous 50 Ω transmission line (TL) with a gap of 0.3 mm, having dimensions of 50 mm × 50 mm on a 1.52 mm thick Roger's TMM4 substrate. In addition, 50 mm × 50 mm dimensions of the substrate and length of 50 mm for TL were selected to ensure the EM wave propagation over the frequency of interest and to compare the size of proposed switches with respect to the discontinuous TL, whereas the width of the TL was calculated using standard impedance matching formulas to match with a 50 Ω SMA connector. The dimensions of the fabricated discontinuous TL are shown in Figure 4. The manufactured MRSs shown in Figure 1d,e were placed on one side of the discontinuous TL. The top side of the MRSs was connected to the other TL using copper tape, as shown in the expanded view in Figure 4. The attached MRSs cavities were filled with an estimated amount of silver coated magnetic particles using the measuring cup procedure explained above. After filling the cavities with the magnetic particles, the top of the MRSs was covered with copper foil and the copper tape was used to connect the top layer with the other end of the discontinuous TL. The photographs of the fabricated discontinuous 50 Ω transmission lines (TLs) with the MRSs attached along with the copper tape connection are shown in Figure 5.

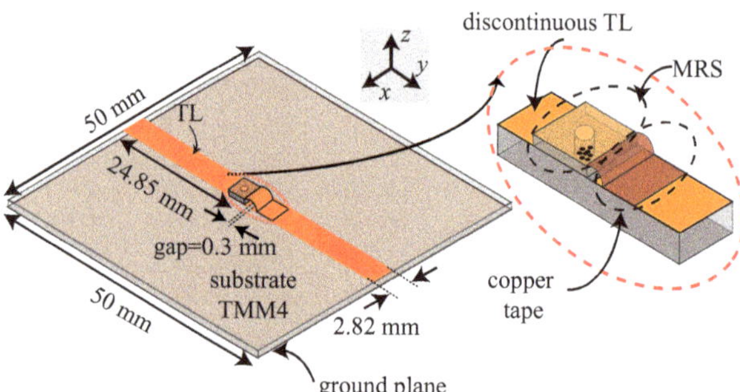

Figure 4. Dimensions of a 50 Ω TL with MRS designed for the S-parameters measurements and extraction of MRS insertion loss and transmission characteristics' results.

The S-parameters, $S_{overall}$ (magnitude and phase), of the manufactured prototype shown in Figure 5a,b were measured using a calibrated network analyzer E5071C in the absence and presence of a magnetic field. A TRL calibration was also performed prior to the measurements to avoid any inaccuracies. Mathematically,

$$S_{overall} = \begin{pmatrix} S_{11_{overall}} & S_{12_{overall}} \\ S_{21_{overall}} & S_{22_{overall}} \end{pmatrix}. \tag{2}$$

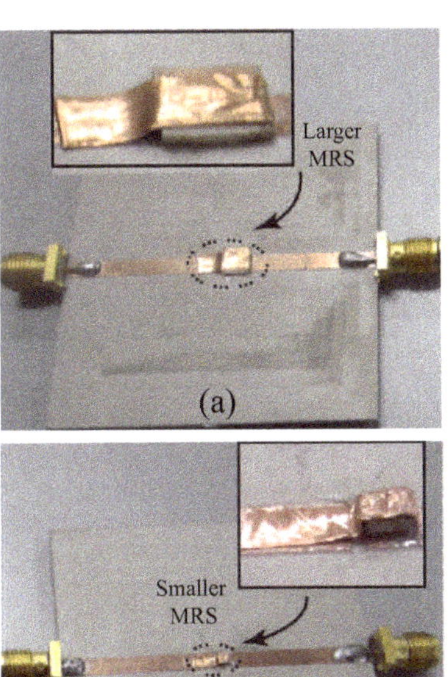

Figure 5. (**a**) MRS shown in Figure 1d attached to a discontinuous 50 Ω TL and (**b**) MRS shown in Figure 1e attached to a discontinuous 50 Ω TL.

These measured S-parameters were then converted into ABCD parameters using standard conversion formulas [15]. For a 50 Ω TL matched with 50 Ω load (Z_{out}) and source (Z_{in}) impedances (model shown in Figure 5), the conversion formulas are:

$$A_{overall} = \frac{\left(Z_{in}^* + S_{11_{overall}} Z_{in}\right)\left(1 - S_{22_{overall}}\right) + S_{12_{overall}} S_{21_{overall}}}{2 S_{21_{overall}} \left(Z_{in} Z_{out}\right)^{1/2}} \tag{3}$$

$$B_{overall} = \frac{\left(Z_{in}^* + S_{11_{overall}} Z_{in}\right)\left(Z_{out}^* + S_{22_{overall}} Z_{out}\right) - S_{12_{overall}} S_{21_{overall}} Z_{in} Z_{out}}{2 S_{21_{overall}} \left(Z_{in} Z_{out}\right)^{1/2}} \tag{4}$$

$$C_{overall} = \frac{\left(1 - S_{11_{overall}}\right)\left(1 - S_{22_{overall}}\right) - S_{12_{overall}} S_{21_{overall}}}{2 S_{21_{overall}} \left(Z_{in} Z_{out}\right)^{1/2}} \tag{5}$$

$$D_{overall} = \frac{\left(1 - S_{11_{overall}}\right)\left(Z_{out}^* + S_{22_{overall}} Z_{out}\right) + S_{12_{overall}} S_{21_{overall}} Z_{out}}{2 S_{21_{overall}} \left(Z_{in} Z_{out}\right)^{1/2}}. \tag{6}$$

where * denotes the complex conjugate. Equations (3)–(6) are used to convert measured S-parameters into ABCD parameters and then arranged in a matrix form as:

$$ABCD_{overall} = \begin{pmatrix} A_{overall} & B_{overall} \\ C_{overall} & D_{overall} \end{pmatrix}. \tag{7}$$

In a separate experiment, two half 50 Ω transmission lines (TLs) having a length of 24.85 mm ($\approx \dfrac{\lambda_\circ}{3}$ @ 3.5 GHz) each were also manufactured and the S-parameters, $S_{left\ half}$ and $S_{right\ half}$ of the half TLs were measured and converted into $ABCD_{left}$ and $ABCD_{right}$ parameters using Equations (3) to (6), respectively. After the extraction of ABCD parameters, the ABCD property was used to extract the ABCD parameters of the MRSs in the presence and absence of a magnetic field, as shown in:

$$ABCD_{overall} = ABCD_{left} \times ABCD_{MRS} \times ABCD_{right} \tag{8}$$

and

$$ABCD_{left}^{-1} \times ABCD_{overall} \times ABCD_{right}^{-1} = ABCD_{MRS}. \tag{9}$$

After the computation of $ABCD_{MRS}$, the insertion loss of the MRSs in the presence of the magnetic field and isolation of the MRSs in the absence of the magnetic field were computed using the standard conversions formulas from ABCD to S-parameters [15]. Finally, the $|S_{21}|$ (dB) values were extracted from the $ABCD_{MRS}$ parameters to observe the insertion losses and isolation performances of the MRSs.

4. Characterization of the Magnetostatic Responsive Structures (MRSs) Using an Equivalent Circuit Model

Next, an equivalent lumped element model of the MRSs was extracted using transmission line theory and microstrip discontinuity theory for a better understanding of the RF switching mechanism and characterization of the MRSs in the absence and presence of a magnetic field. A side view of the MRS and copper tape along with the proposed equivalent lumped element model on a host 50 Ω TL is shown in Figure 6. The gap discontinuity between the two open ended transmission lines was modeled using the gap discontinuity theory [16]. The capacitance C_f was introduced due to the fringing effect of the open ends of the TL and a slight extension of the electric fields without stopping abruptly, whereas the coupling capacitance of the gap was modeled as C_{gap}. The values of coupling capacitance (C_{gap}) and fringing capacitance (C_f) were calculated using formulas given in [12,17]. Since the bottom of the MRSs was attached to the one end of the discontinuous TL and the top surface of the MRSs was connected using copper tape to the other half of the discontinuous 50 Ω TL, the copper tape was considered as a suspended microstrip line of finite length over the air and the host TL substrate (TMM4). Thus, the suspended portion of the copper tape was modeled as a series inductance, L_{tape} along with the capacitances; C_{tape}° and C_{tape}^{d}. The capacitances were attributed to the suspended portion of tape over air and dielectric. C_{tape}° represents capacitance due to the air, whereas capacitance due to dielectric is represented by C_{tape}^{d}. These values were calculated using the following [16]:

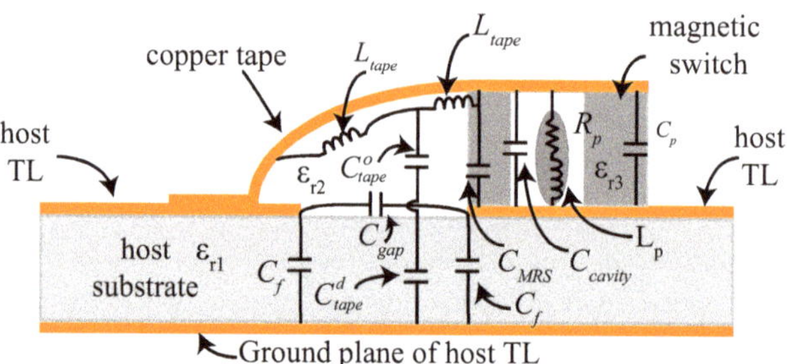

Figure 6. Side view of MRS attached to the host 50 Ω TL with a proposed equivalent lumped element model.

$$\frac{L_{tape}}{h}(nH/m) = 100\left\{4\sqrt{\frac{W}{h}} - 4.21\right\} \tag{10}$$

and

$$C_{tape}^d(pF/m) = \frac{(14\epsilon_r + 12.5)\, W/h}{\sqrt{W/h}} - \frac{(1.83\epsilon_r + 2.25)}{\sqrt{W/h}} + \frac{0.02\epsilon_r}{W/h}$$
$$= C_{tape}^{\circ}(pF/m) \tag{11}$$

when $W/h < 1$. For $W/h \geq 1$, equations for C_{tape}° and C_{tape}^d will be

$$C_{tape}^d(pF/m) = (9.5\epsilon_r + 1.25)\, W/h - 5.2\epsilon_r + 7.0$$
$$= C_{tape}^{\circ}(pF/m), \tag{12}$$

where ϵ_r is the relative permittivity of the substrate, W is the width of the transmission line (which is the same as that of copper tape), and h is the height of the substrate and air gap between the suspended copper tape and ground plane.

Moreover, shunt capacitances C_{MRS} were accounted for because of the embodiment of the structure, whereas C_{cavity} resulted from the cavity between the top and bottom copper tape. In addition, the capacitance C_p accounted for the effect of the open-end fringing capacitance of the open-end TL and calculated using the closed form expressions given in [12,16]. The values of C_{tape}°, and C_{tape}^d were estimated using Equations (7) and (8), whereas C_{MRS} and C_{cavity} were calculated using the formula for parallel plate capacitance [15]:

$$C_{MRS}(F) = \epsilon_0\epsilon_r W/h = C_{cavity}(F) \tag{13}$$

Furthermore, the resistance of the magnetic particles $R_{particle}$ was added in the proposed equivalent circuit model of the MRSs to encounter the absence and presence of the static magnetic field. The value of the $R_{particle}$ was made very high because of the air gap in the cavity between and the top copper tape and bottom conductor, as silver coated magnetic particles settled on the bottom surface in the absence of the magnetic field. However, a smaller value of $R_{particle}$ was chosen when a magnetic field was applied to the MRSs.

On the other hand, the presence of the magnetic field aligned the magnetic particles and provided a current path between the top and bottom conductors [17,18]. Since the quantity of the magnetic particles was estimated using the measuring cup procedure and Equation (1), the minimum number of magnetic particles per column N_s were estimated using the following formula:

$$N_s = \frac{h}{d}, \tag{14}$$

where h is the substrate height of the MRSs and d is the diameter of the magnetic particles. Then, the total number of columns N_c in the MRS cavity was estimated using:

$$N_c = \frac{N_t}{N_s}, \tag{15}$$

where N_t is the total number of magnetic particles in a cavity and N_s is the minimum number of magnetic particles per column. It was concluded that, in the presence of the magnetic field, silver coated magnetic particles built N_c columns with N_c paths between the top copper and bottom conductor.

These paths were considered as a conducting via of N_c columns having a radius of $N_c \times$ average diameter of magnetic particles, i.e., 40 µm. Therefore, the inductance of the conducting via was modeled as L_p and extracted using the theory of microstrip vias [18]. The formula used for estimating the inductance of via L_p formed by columns of magnetic particles in the MRS cavity was:

$$L_p = \frac{\mu_0}{2\pi}\left[h\ln\left(h\ln\left(\frac{2h + \sqrt{r^2 + h^2}}{r} \right) + \frac{3}{2}\left(r - \sqrt{r^2 + h^2} \right) \right) \right], \tag{16}$$

where r is the average radius of the via formed by the magnetic particle columns present in the MRS cavity and h is the height of the MRS substrate.

In the absence of the magnetic field, the magnetic particle did not build up columns in the MRSs cavity, but there were still magnetic particles settled down in the bottom conductor. Thus, for the absence of the magnetic field, the radius r and height h in Equation (16) are equal to the radius of the measuring cup used to fill the MRSs cavity. A complete proposed equivalent lumped element model of the MRSs is shown in Figure 6.

5. Results and Discussion

The insertion loss of the smaller and larger MRSs shown in Figure 1d,e, respectively, were extracted using the ABCD parameters method and procedure shown in Equations (2)–(9). The insertion loss and isolation characteristics of both the MRSs are shown in Figure 7. It can be seen from Figure 7a that both of the MRSs have an insertion loss ($dB(|S_{21}|)$) less than 1.5 dB in the presence of the magnetic field i.e., in the 'ON' state. However, the smaller MRS has a lower insertion loss, but the general behavior of the insertion loss was the same for both the MRSs. At lower frequencies, the insertion loss was less than 0.5 dB, whereas the insertion loss up to 1.5 dB was observed for higher frequencies. Overall, both the MRSs showed good insertion loss, and it was concluded that the novel structures have a capability to propagate an EM wave over the range of frequencies from 100 kHz–3.5 GHz in the 'ON' state.

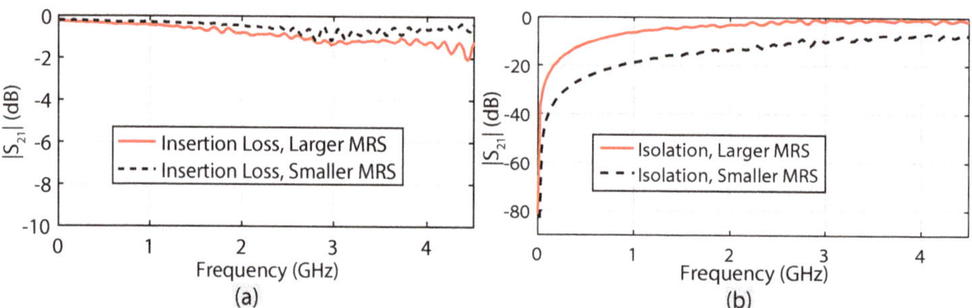

Figure 7. Insertion loss and Isolation comparison of the larger and smaller MRSs in the absence and presence of the static magnetic field using ABCD parameters method. (**a**) Insertion Loss and (**b**) Isolation characteristics.

The isolation characteristics of the MRSs in the absence of the static magnetic field i.e., in 'OFF' state were also extracted using the ABCD parameters method and are depicted in Figure 7b. It was observed that an isolation of 10 dB up to several 100 MHz for the larger MRS was achieved, whereas better isolation performance up to 3.5 GHz was observed in the smaller MRS. The S-parameters extraction of the MRSs from the ABCD parameters and conversion into insertion and isolation performance showed that these novel structures have the potential to be used in RF switching over the frequency range from 100 kHz to 3.5 GHz.

Next, for the validation of the MRSs' characterization results from ABCD parameters, the values of the proposed equivalent lumped element circuit shown in Figure 6 for the smaller MRS were extracted using the equations and lumped elements theory discussed in Section 4. Figure 8 shows the extracted values of the lumped elements for the smaller MRS in the absence of the static magnetic field, whereas all the extracted values remained unchanged except for the $R_{particle}$ = 0.2 Ω and L_p in the presence of a magnetic field.

The circuit shown in Figure 8 was simulated in Keysight Advanced Design Software (ADS) v. 15.0 [19] to observe the insertion loss and isolation performance of the smaller MRSs. In simulations, the presence and absence of the magnetic field were mimicked by changing the values of $R_{particle}$ and L_p. A comparison between the insertion loss and isolation extracted from the circuit model and ABCD parameters method is shown in Figure 9. It was observed that the insertion loss of the MRSs was less than 1.5 dB over a frequency range from 100 kHz to 3.5 GHz, whereas an isolation up to 10 dB was achieved. The results comparison of the insertion and isolation values from the ABCD parameters validated the equivalent circuit model and vice versa. Moreover, the results comparison in Figure 9 showed that the values of the equivalent lumped element in Figure 6 can easily be extracted for any geometry of the MRSs having different dimensions to achieve better insertion and isolation performances at higher frequencies i.e., greater than 3.5 GHz. It can also be observed from Figure 9b that the isolation results from the equivalent circuit model are in good agreement with the results extracted from the ABCD parameters method.

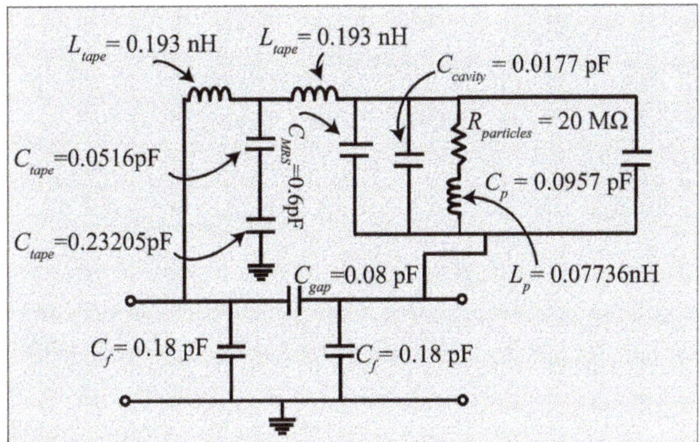

Figure 8. Lumped elements values of the smaller MRS in the absence of magnetic field.

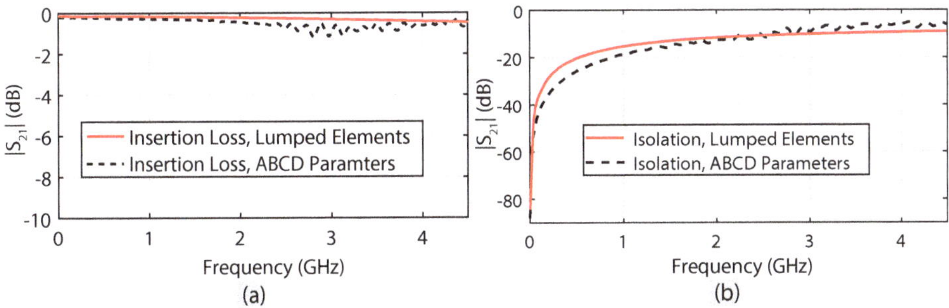

Figure 9. Comparison of insertion loss and isolation of smaller MRS with extracted insertion loss and isolation characteristics from the equivalent lumped element model and ABCD parameters method. (**a**) Insertion Loss and (**b**) Isolation characteristics.

Furthermore, to demonstrate the working of the proposed MRSs as a RF switch, a rectangular microstrip patch antenna is first designed using standard formulas and a small rectangular strip of

3.3 mm width is placed at a distance of 1.7 mm from the rectangular patch [11]. The small rectangular strip is connected with the main portion of rectangular patch using the proposed smaller MRSs, as shown in Figure 10a. To mimic the 'ON' state of MRSs in simulation, conducting vias connecting the top and bottom that was the main rectangular patch were modelled, whereas, conducting vias were removed in the simulation setup to mimic the 'OFF' state of MRSs. On the other hand, during the measurement of magnitude of reflection coefficient ($|S_{11}|$ (dB)), a permanent magnet was used to switch the 'OFF' and 'ON' state of the MRSs. A comparison of simulation and measurements results is shown in Figure 10b. It can be observed from Figure 10b that 'OFF' state of the MRSs disconnected the main patch from the small rectangular strip and antenna resonated at 2.42 GHz. On the contrary, 'ON' state of the MRSs provided a connection between both the patches and increased overall electrical length of the resonating structure, thereby resulting in a lower resonant band of 2.1 GHz. In total, 10 smaller MRSs were used in the demonstration process to provide a continuous current path between the two conducting patches in 'ON' state of the MRSs.

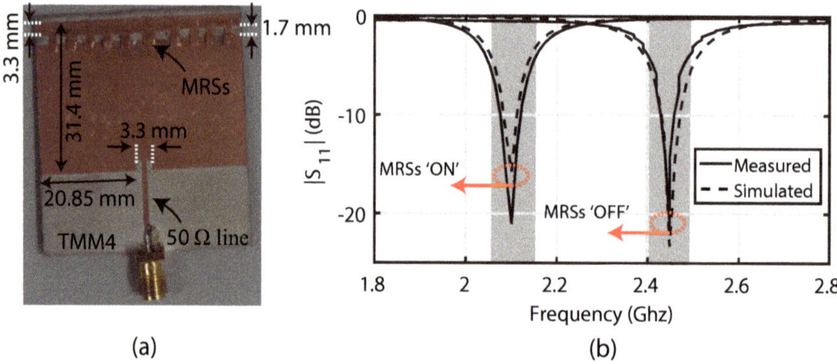

Figure 10. (**a**) fabricated prototype of reconfigurable microstrip patch antenna with the attached MRSs and (**b**) comparison of simulated and measured $|S_{11}|$ (dB).

It can be observed from the RF characterization results of the proposed MRSs that these switches can be used in reconfigurable antennas operating from 100 kHz to 3.5 GHz. The proposed RF switches have better insertion loss of −1.5 dB and good isolation performance. However, one may note the high isolation required by reconfigurable antennas. Therefore, increasing the isolation of RF switches beyond −10 dB is the main bottleneck of these switches for high isolation antenna applications. Another limitation of these RF switches can be their operation during rotation along with their possible use in higher band application beyond 3.5 GHz. These limitations can be addressed using symmetric and asymmetric copper files along with the use of thicker embodiment of the RF switches and using multiple holes filled with magnetic particles in the embodiment [20].

The RF characterization and realization procedure of the novel structures carried out in this work reveal that these structures have a plethora of applications in reconfigurable and tunable antenna design to provide multi-functionality. Another application avenue of the proposed RF switches can be in the fast near-field measurements systems used in microwave imaging with the use of a Modulated Scattering Technique (MST) [21]. The array of probe antennas loaded with RF PIN diodes can be replaced with the proposed novel switches to measure the near-field distribution over the surface. In addition, the embedment feature of the proposed MRS can be very useful in antennas used for real-time near field measurements with multi-probe technologies under control topology of optimization algorithms and proper switching of MRSs.

6. Conclusions

This paper presented the development of micro-level MRSs. A procedure of extracting insertion loss and isolation characteristics of novel MRSs using ABCD parameters has also been presented along with the proposed equivalent circuit model using the theory of microstrip lines. It was successfully shown that the proposed novel structures consisting of the magnetostatic responsive structures have good RF switching responses from 100 kHz to 3.5 GHz. Therefore, it can be concluded that the smaller structure has better switching capabilities up to 3.5 GHz, whereas the larger MRSs can only perform up to 700 MHz. Additionally, the realization and characterization of the novel MRSs revealed that these structures have the capability to be embedded in the RF instrumentation substrate. Once properly packaged, these structures can be an alternative to the existing RF switches such as PIN diodes and FETs.

Author Contributions: B.D.B. provided the idea. A.I., S.M.A., J.M.P., and A.F. performed complete simulations, fabrications, measurements, and post processing. J.A., M.A., and D.E.A. assisted in overall paper management, idea development, and manuscript writing. All authors have read and agreed to the published version of the manuscript.

Funding: This work was supported by the US Air Force Research Laboratory Munitions Directorate under Grant No. FA-8651-15-2-002.

Conflicts of Interest: The authors declare no conflict of interest.

References

1. Bacon, P.; Fischer, D.; Lourens, R. Overview of RF switch technology and applications. *Microw. J.* **2014**, *57*, 76–88.
2. Skyworks Solutions Inc. Choosing the Right RF Switches for Smart Mobile Device Applications (White Paper). Available online: www.skyworksinc.com (accessed on 15 September 2019).
3. Anagnostou, D.E.; Guizhen, Z.; Chryssomallis, M.T.; Lyke, J.C.; Ponchak, G.E.; Papapolymerou, J.; Christodoulou, C.G. Design, fabrication, and measurements of an RF-MEMS-based self-similar reconfigurable antenna. *IEEE Trans. Antennas Propag.* **2006**, *54*, 422–432. [CrossRef]
4. *Reed Switch-KSK-1A04*; Meder Electronic: Singen, Germany, 2015. Available online: www.standexmeder.com (accessed on 15 September 2019).
5. Wu, C.; Wang, T.; Ren, A.; Michelson, D.G. Implementation of Reconfigurable Patch Antennas using Reed Switches. *IEEE Antennas Wirel. Propag. Lett.* **2011**, *10*, 1023–1026.
6. Liu, Q.; Wang, N.; Zeng, Q.; Wu, C.; Wei, G. A Frequency Reconfigurable Patch Antenna with Reed Switches. In Proceedings of the 2014 IEEE International Wireless Symposium (IWS 2014), Xi'an, China, 24–25 March 2014.
7. Potters Industries LLC. Available online: http://www.pottersbeads.com/ (accessed on 15 September 2019).
8. Rogers Corporation. Available online: www.rogerscorp.com (accessed on 15 September 2019).
9. Iftikhar, A.; Parrow, J.M.; Asif, S.M.; Sajal, S.Z.; Braaten, B.D.; Allen J.; Allen, M.; Wenner, B. A printed dipole reconfigured with magneto-static responsive structures that do not require a directly connected biasing circuit. In Proceedings of the 2016 IEEE International Symposium on Antennas and Propagation (APSURSI), Fajardo, Puerto Rico, 26 June–1 July 2016; pp. 1057–1058.
10. Iftikhar, A.; Parrow, J.M.; Asif, S.M.; Braaten, B.D.; Allen, J.; Allen, M.; Wenner, B. On using magneto-static responsive particles as switching elements to reconfigure microwave filters. In Proceedings of the 2016 IEEE International Conference on Electro Information Technology (EIT), Grand Forks, ND, USA, 19–21 May 2016; pp. 192–195.
11. Iftikhar, A.; Parrow, J.M.; Asif, S.; Allen, J.; Allen, M.; Werner, B.; Braaten, B.D. Improving the efficiency of a reconfigurable microstrip patch using magneto-static field responsive structures. *IET Electron. Lett.* **2016**, *52*, 1194–1196. [CrossRef]
12. Parrow, J.M.; Iftikhar, A.; Asif, S.M.; Allen, J.; Allen, M.; Werner, B.; Braaten, B.D. On the bandwidth of a microparticle-based component responsive to magneto-static fields. *IEEE Trans. Electromagn. Compat.* **2017**, *59*, 1053–1059. [CrossRef]
13. LPKF Protomat S103. Available online: www.lpkf.com (accessed on 15 September 2019).

14. Hsiang, W.Y. On the sphere packing problem and the proof of Kepler's problem conjecture, *Int. J. Math.* **1993**, *4*, 739–780.

15. Pozar, D.M. *Microwave Engineering*, 4th ed.; John Wiley and Sons: Hoboken, NJ, USA, 2011; pp. 165–227.

16. Garg, R.; Bahl, I.; Bozzi, M. *Microstrip Lines and Slotlines*, 3rd ed.; Artech House: Norwood, MA, USA, 1996.

17. Swanson, D.G.; Hoefer, W.J.R. *Microwave Circuit Modeling Using Electromagnetic Field Simulation*; Artech House: Norwood, MA, USA, 2003.

18. Goldfrab, M.E.; Pucel, R.A. Modeling via hole grounds in microstrip. *IEEE Microw. Guid. Wave Lett.* **1991**, *1*, 135–137.

19. Keysight Technologies, Advanced Designed System (ADS). Available online: www.keysight.com (accessed on 15 September 2019).

20. Balenji, N.S. A New Generation of Particle-Based Radio Frequency (RF) Switches for Portable and High-Frequency Applications. Master's Thesis, North Dakota State University, Fargo, ND, USA, 2019.

21. Bolomey, J.C.; Gardiol, F. *Engineering Applications of the Modulated Scattering Technique*; Arctech House Press: Norwood, MA, USA, 2010.

Article

A Reconfigurable Polarization—Frequency Supershape Patch Antenna with Enhanced Bandwidth

Anastasios Koutinos, **Georgia Xanthopoulou**, **George Kyriacou** and **Michael Chryssomallis** *

Democritus University of Thrace, Electrical & Computer Engineering Department, 67100 Xanthi, Greece; akoutino@ee.duth.gr (A.K.); georxant8@ee.duth.gr (G.X.); gkyriac@ee.duth.gr (G.K.)
* Correspondence: mchrysso@ee.duth.gr

Received: 1 May 2020; Accepted: 15 July 2020; Published: 18 July 2020

Abstract: In this article a reconfigurable antenna for WLAN/WiMAX applications is presented. A super-shape radiator of an ellipsis shape is used to achieve wider intrinsic bandwidth compared to the classical rectangular patch antenna, while the dimensions remain comparable. The proposed antenna is fed at two points exciting both horizontal and vertical polarization but in different operating frequencies. To achieve wider bandwidth, as a whole but also for each polarization, the symmetrical feeding points for each excitation are also employed with a proper feeding network. PIN diodes are also used in the feeding network to provide the option of narrower bandwidth. The antenna substrate is Rogers RO4003C with dielectric constant $\varepsilon_r = 3.55$ and dissipation losses $\tan\delta = 0.0027$ with height h = 1.524 mm. The antenna operates in the range of 2.3 GHz to 2.55 GHz but, using the proposed procedure, it can be designed for different frequency ranges.

Keywords: wideband reconfigurable antennas; new feeding techniques; dual polarization; multifrequency antennas; supershape

1. Introduction

Modern day wireless communication systems have radiation requirements that can be satisfied using different antennas. In order to achieve a more compact structure and if possible a single device, reconfigurable antennas are one of the solutions that are able to meet these expectations [1]. Frequency, polarization and radiation patterns are some of the antenna properties that would be required to be reconfigurable if a multifunction device would be designed. From the characteristics mentioned above, the radiation pattern is of minor importance because in general it can be achieved through phased arrays. This is achieved in [2] with the implementation of a reconfigurable feeding network, and in [3], through the use of a reconfigurable frequency selective surface (FSS). A different approach to have radiation pattern reconfigurability is to actually design an antenna, which has an omnidirectional pattern consisting of several identical endfire sectors pointing to opposite directions. In this way, by activating a different part of the antenna, a corresponding pattern is enabled, resulting in pattern-reconfigurable properties. This is properly explained in [4], where the omnidirectional pattern contains two identical patterns, while in [5], the same principle is studied but with four brackets for the antenna and four patterns as well. Another important characteristic of reconfigurable antennas is the possibility to operate on different frequencies. This is presented in [6], where a wideband antenna with a monopole topology is studied. Between the radiating element and the feeding port, a reconfigurable bandpass filter (BPF) is placed which has the ability to shift its central frequency and so the whole structure can alter its resonant frequency. Actually, this antenna can operate to wideband or narrowband states with the narrowband one offering several adjacent frequencies one at a time. An alternate method for frequency reconfigurability is shown in [7], where through the activation of different combinations of PIN diodes, adjacent frequencies can be also utilized. The PIN diodes can also be applied to achieve frequency altering in case of multiband behavior as explained in [8]. Here the only striking diversity is that while the general resonant properties remains the same, each case of PIN diode configuration result to better matching for different frequencies of the same bandwidth.

In a similar way as described in [6] wideband antennas can be designed. The main principle is that one can utilize an UWB antenna and with the proper modifications, to exclude some frequency bands and thus a wideband behavior can be achieved. This is presented in [9], where the UWB antenna is a Bow-Tie in monopole topology. Using a number of switches three states can be studied, each with its own central frequency and bandwidth (BW) as well. The same principle is exhibited in [10] with the only difference being that the distinct states are able to cover the entire BW with one of them being redundant (OFF-ON). Another example of this procedure is presented in [11] where two "nearly" monopoles, mainly because the ground plane is not entirely absent for them, are used. Here their ground plane consists of a reconfigurable branched structure with which some bands are forfeited as seen by the gain and efficiency figures. The major drawback of the monopole structures is the low efficiency due to their omnidirectional pattern and specifically the existence of a back-lobe.

Finally, for an antenna to become reconfigurable is in its polarization. In [12], two monopoles are used using switches to render each one active or not. Each monopole has linear polarization (LP) and they are fed by the same port. Since they are placed vertically, three possible linear (horizontal, vertical and their combination aka diagonal) polarizations are observed. A similar idea is presented in [13], but here, a slot array antenna is used with each port having LP and thus dual LP is achieved. The concept of dual LP is studied in [14], employing two vertical dipoles each with its own LP as well. Another approximation on the polarization issue is the exploration of right hand circular polarization or left hand circular polarization (RHCP and LHCP) which is carried out in [15]. Here using two switches, the current distribution direction on the radiating element can be controlled and thus RHCP or LHCP can be realized but with the same radiation pattern. Related to [15] but with more complexity is the antenna proposed in [16] where RHCP or LHCP can be realized on demand in a fairly wide BW.

The issue we try to address is to have an antenna with different polarizations which is operating within a fair BW but also with ease of fabrication as well as keeping the antenna as compact as possible. Most of the approaches from literature have either a large number of PIN diodes or bigger dimensions. Therefore, at this paper an ellipsis Supershape Patch Antenna (SPA) is proposed. A SPA has the same bandwidth compared to a rectangular operating at the same central frequency but utilizes circular polarization and so common radiation properties can be achieved through the entire operation bandwidth. Our work can be directly compared to that of [17], where a rectangular patch, with its intrinsic narrow BW and linear polarization, was designed and one can see that a similar design procedure is carried out, resulting to significantly narrower bandwidth for the circular polarization.

2. Antenna Design

2.1. Dual Fed SPA

In order to utilize both resonances (modes TM_{10} and TM_{10}) for greater bandwidth, our initial approach will start using a rectangular patch. The major drawback however is that different modes radiate with horizontal, and vertical polarization, respectively, thereby making the entire bandwidth unusable as a whole. To counter this issue, an ellipsis SPA is chosen mainly because Supershape geometries can be used as antenna structures with similar characteristics as studied in [18]. The main feature of Supershape antennas is the circular polarization even when they are fed at a single point. The dielectric substrate was chosen to be Rogers RO4003C with dielectric constant $\varepsilon_r = 3.55$, dissipation losses $\tan\delta = 0.0027$ and thickness h = 1.0524 mm. The geometry shape of the ellipsis is a result of Equation (1), known as Supershape formula, taken from [19]. Following up, the ellipsis is scaled by a factor of 1.271 for the vertical dimension. The scaling is done in order the resonant frequency of the vertical side to shift appropriately and come closer to the one of the horizontal side. After the scaling the geometry looks like a circle but this is because the two axes (r_1 and r_2) are almost but not exactly equal.

$$r(\varphi) = \left[\left| \frac{1}{a} cos\left(\frac{m}{4} \varphi \right) \right|^{n_2} + \left| \frac{1}{b} sin\left(\frac{m}{4} \varphi \right) \right|^{n_3} \right]^{-n_1} \tag{1}$$

In Equation (1), terms a and b represent the horizontal and vertical dimension of the resulting shape. The "contribution" of each term is determined by n_2 and n_3 while n_1 defines the sharpness of the supershape corners. As stated in [19], "For $n_1 = n_2 = n_3 = 2$ and $m = 4$ an ellipse is obtained. A circle is obtained when additionally, $a = b$". The Equation (1) has no units because the supershape dimensions can be assigned to any (i.e. mm or cm etc). To achieve dual band behavior, two feeding points are chosen for the proposed SPA, one on each of the two axes. The geometry of the antenna, with the positions of feeding points, is shown in Figure 1.

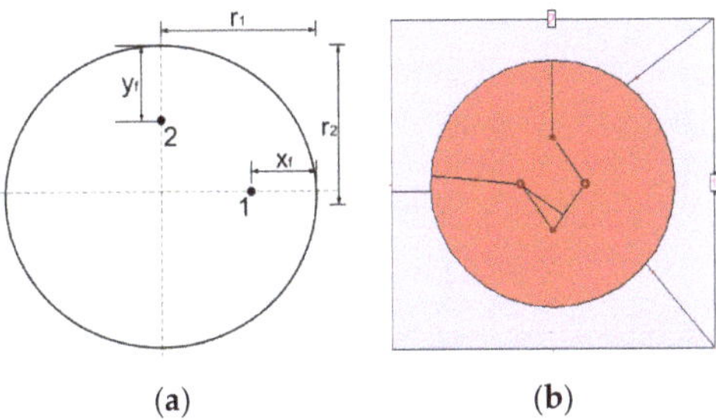

(a) (b)

Figure 1. Supershape Patch antenna geometry (**a**) Dimension design and (**b**) IE3D design.

The positions of feeding points #1 and #2 are chosen to have different impedance in order to spare space for the placement of the feeding network. Due to the geometry of the antenna, the horizontal side has smaller input impedance value than the vertical one on the edge. As such, if both feeding points were chosen to have $Z_o = 50$ Ohm, the feeding ports distance would be small enough and coupling phenomena would occur between them with the result of poor radiation. The parameters of the antenna are given in Table 1 as well as the different port impedances.

Table 1. Values of Supershape patch parameters.

Parameter	Value	Parameter	Value
m	1	r_1	19 mm
n_1	0.5	r_2	18.71 mm
n_2	5	x_f	14 mm
n_3	5	y_f	11.05 mm
a	0.8902	Z_{01}	50 Ohm
b	0.8902	Z_{02}	130 Ohm

The simulations have been carried out using the Method of Moments (MoM) implemented by the Zeland IE3D simulator. For the characterization of the antenna, the S-parameters are used which provide information about the well matching of the SPA. As expected, each feeding point excites a single resonance while the coupling between the two ports is insignificant (below −25 dB) as presented in Figure 2.

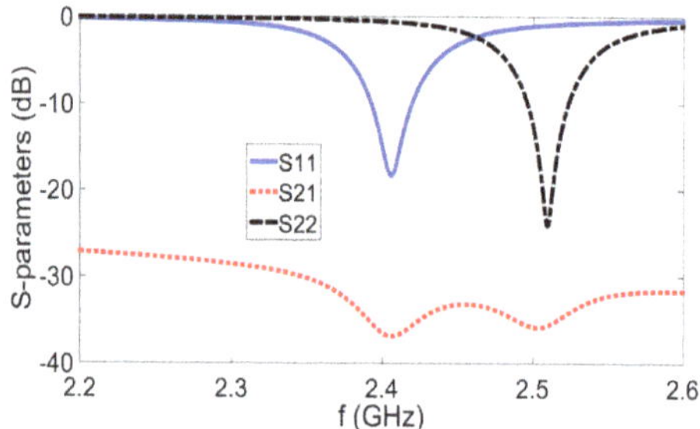

Figure 2. S–Parameters of supershape patch antenna.

As seen in Figure 2 each resonance is well matched (below −10 dB). In Table 2 all the important frequencies of the SPA are given along with the bandwidth of each resonance. The two resonances are chosen with no common frequencies because the bandwidth of each resonance is to be enhanced and so a considerable wider bandwidth is formed.

Table 2. Important frequencies of the SPA.

	Resonance 1 (S_{11})	**Resonance 2 (S_{22})**
f_L (GHz)	2.4006	2.5033
f_H (GHz)	2.4366	2.5314
f_C (GHz)	2.4186	2.5173
BW (MHz)	36	28.1
BW (%)	1.48	1.11

2.2. Bandwidth Enhancement of Dual Fed SPA

The SPA is modified to increase its bandwidth. The bandwidth enhancement is implemented using a matching feeding network as described in [20] and in a similar way to the one in [17]. The feeding points 1, 2 and 3, 4 are connected through separate feeding networks for 50 Ohm and 130 Ohm respectively due to lack of space as well as to avoid coupling effects between the ports. The proposed geometry is multilayer having the radiating patch on the top layer, the middle layer serving as ground plane while the feeding network is to be placed on the bottom layer. This can be easily done using different substrates for the patch and the feeding network and glue them together to keep common ground plane afterwards. The different layers of the geometry of the proposed modified SPA are shown in Figure 3 and its dimensions are given in Table 3.

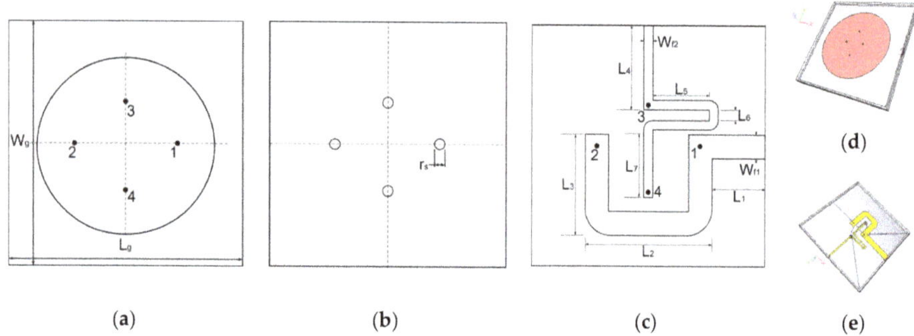

Figure 3. The three layers of the geometry of the proposed modified SPA, (**a**) patch (top), (**b**) ground plane (middle), (**c**) feeding network (bottom), (**d**) IE3D top view and (**e**) IE3D bottom view.

Table 3. Modified SPA dimensions (all values in mm).

Parameter	Value	Parameter	Value
L_g	50	L_2	13.5
W_g	50	L_3	15.45
r_s	1	L_4	18.4
W_{f1}	3.5	L_5	9.9
W_{f2}	0.8	L_6	1.8
L_1	18.25	L_7	11.8

The size of the proposed antenna is kept the same (ground plane dimensions) but its thickness is doubled because of the second dielectric substrate. The antenna is fed from both ports simultaneously to confirm the bandwidth increase of each resonance but also the bridging of the gap between the two separate resonances, so the S-parameters of the antenna are given in Figure 4.

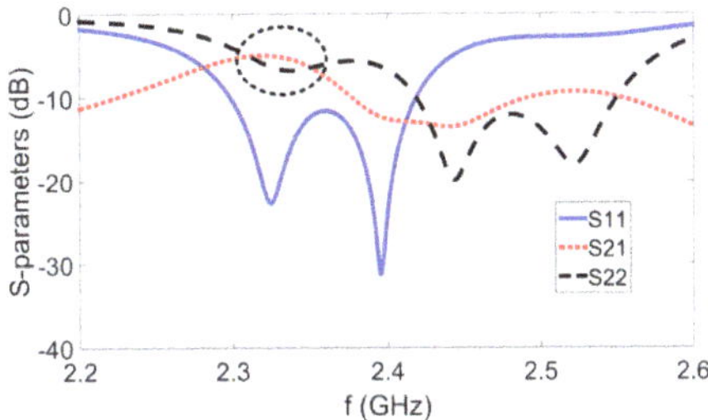

Figure 4. S–Parameters of modified SPA.

As it is observed, there is an issue with the coupling between the two ports because it is not negligible as it is for the SPA (Figure 2). Especially in 2.32 GHz (black circle) S_{21} has its maximum value (≈-5 dB) and as such it is expected to have little to none radiation to that frequency area. This could be countered using a proper matching circuit, which will transform the Z_{in2} = 130 Ohm to 50 Ohm while will keep the Z_{in1} = 50 Ohm at the same time, as it has been described in [21]. Given that, in Table 4 the frequencies and bandwidth of each resonance are listed.

Table 4. Important frequencies of the SPA.

	Resonance 1 (S_{11})	Resonance 2 (S_{22})
f_L (GHz)	2.2985	2.4187
f_H (GHz)	2.4188	2.5501
f_C (GHz)	2.3586	2.4844
BW (MHz)	120.3	131.4
BW (%)	5.1	5.29

The two resonances are barely overlapping, as shown from the higher frequency of the first resonance and the lower frequency of the second resonance, forming the expected wider bandwidth. The modified SPA is going to be further altered by using PIN diodes in order to achieve the reconfigurable behavior we are looking for.

3. Reconfigurable SPA

The use of PIN diodes can give different radiation and matching characteristics leading to the result that modified SPA becomes a reconfigurable antenna. The antenna with the PIN diodes is shown in Figure 5, where the positions of diodes are shown (the blue rectangles), while the other dimensions are the same as before, given in Table 3. Also, the top and bottom views of the fabricated prototype are given. The different states of the PIN diodes (ON or OFF) create 4 possible cases to explore, as explained in Table 5.

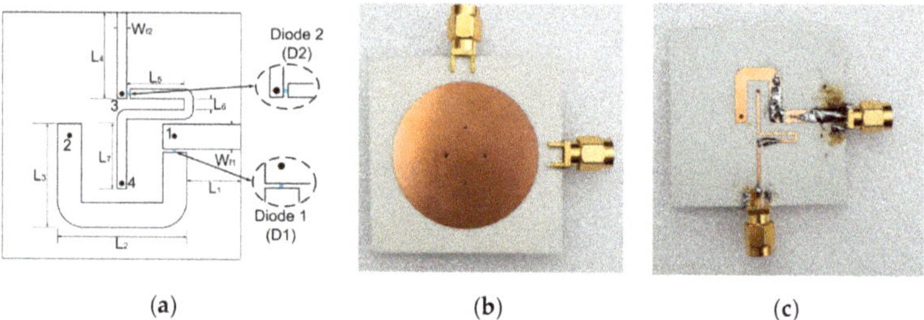

(a)   (b)   (c)

Figure 5. Geometry of SPA (**a**) Bottom side of the modified SPA with PIN diodes, (**b**) Top view of the prototype SPA and (**c**) Bottom vie of the prototype SPA.

Table 5. Different configurations of PIN diodes.

Case	State	
	D1	D2
1	Off	Off
2	On	Off
3	Off	On
4	On	On

For each of these four cases, a different simulation has been carried out in order to determine its radiation properties. The simulated results of the return losses (S_{11}), (S_{22}) and the coupling of the two ports (S_{21}) for each of the cases of Table 5 are given (with the S_{12} not given due to being equal to (S_{21}) in Figure 6. In Table 6 all the frequencies of each case are shown.

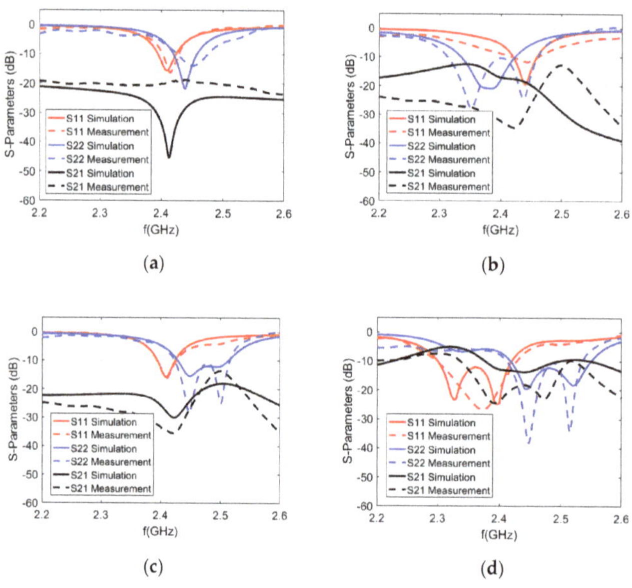

(a)   (b)

(c)   (d)

Figure 6. S—Parameters for each case of Table 5, (**a**) Case 1, (**b**) Case 2, (**c**) Case 3 and (**d**) Case 4.

Table 6. Important frequencies of the SPA for the different cases of Table 5.

	Case 1		Case 2		Case 3	
	Resonance 1 (S_{11})	Resonance 2 (S_{22})	Resonance 1 (S_{11})	Resonance 2 (S_{22})	Resonance 1 (S_{11})	Resonance 2 (S_{22})
f_L (GHz)	2.3946	2.4205	2.3352	2.4202	2.3949	2.4294
f_C (GHz)	2.4095	2.4393	2.3788	2.4392	2.4098	2.4711
f_H (GHz)	2.4245	2.4582	2.4224	2.4582	2.4248	2.5129
BW (MHz)	29.9	37.7	87.2	38	29.9	83.5
BW (%)	1.24	1.54	3.66	1.55	1.24	3.38

As one can easily see, for the three of the four cases (cases 1, 2 and 3) the coupling between ports is very weak ($S_{21} < -10$ dB). Especially in Case 1 and 3, as shown in Figure 6a,c, the coupling is even lower ($S_{21} < -15$ dB) and so the two resonances have good radiation for their respected frequency areas. This is also indicative of the fact that each port could be activated separately and so one of the resonances could be excited any time.

Observing the frequency limits of each resonance in Table 6, it is obvious that only for the Case 2 the resonances are actually overlapping ($f_{Hres1} < f_{Lres2}$). However, for the other two cases, the frequencies have small deviation and so a whole bandwidth consisting of both resonances can be formed. Having the S-parameters of the structure we have the flexibility of exciting either both ports or one of them depending on our demands for smaller or larger bandwidth. Therefore, for each case the different combinations of port excitations are given in Table 7. The Case 4 is not further explored here due to lack of proper matching.

Table 7. Different configurations for Port 1–Port 2.

Case 1	Case 2	Case 3
On-Off	On-Off	On-Off
Off-On	Off-On	Off-On
On-On	On-On	On-On

By nature, the supershape patch has circular-elliptical polarization, so to further study and determine this, the radiation patterns for each configuration of Table 7 are given in Figures 7–9 and in each one the left hand circular polarization (LHCP) and right hand circular polarization (RHCP) are shown. Also the E-total is given, depicting the radiation pattern of the antenna in general.

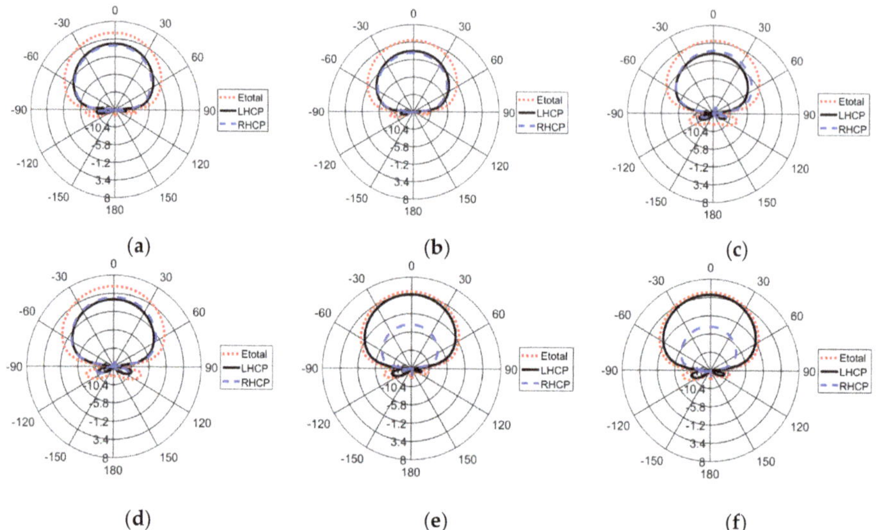

Figure 7. Radiation patterns for center frequencies of each resonance for Case 1 (**a**) f = 2.409 GHz for On-Off, (**b**) f = 2.439 GHz for On-Off, (**c**) f = 2.409 GHz for Off-On, (**d**) f = 2.439 GHz for Off-On, (**e**) f = 2.409 GHz for On-On and (**f**) f = 2.439 GHz for On-On.

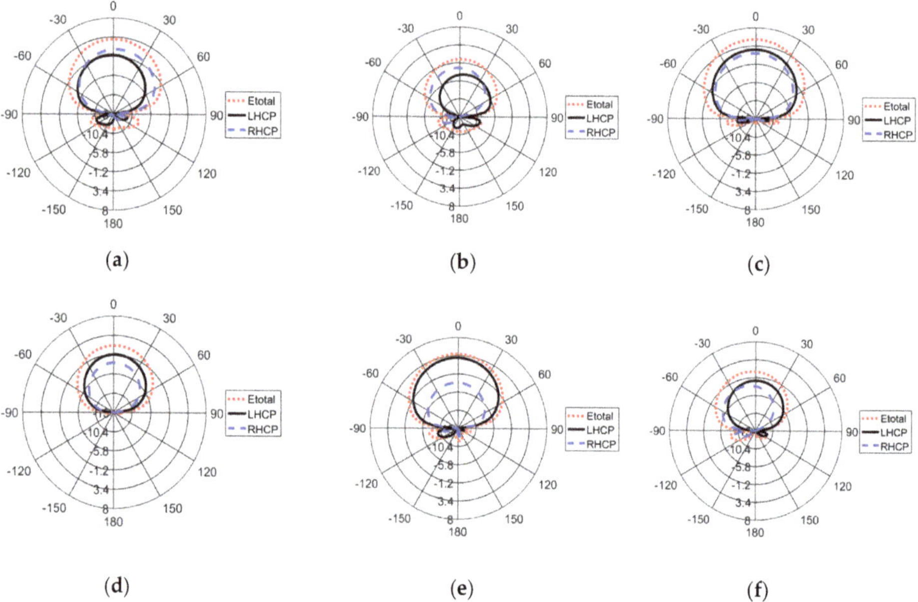

Figure 8. Radiation patterns for center frequencies of each resonance for Case 2 (**a**) f = 2.409 GHz for On-Off, (**b**) f = 2.479 GHz for On-Off, (**c**) f = 2.409 GHz for Off-On, (**d**) f = 2.479 GHz for Off-On, (**e**) f = 2.409 GHz for On-On and (**f**) f = 2.479 GHz for On-On.

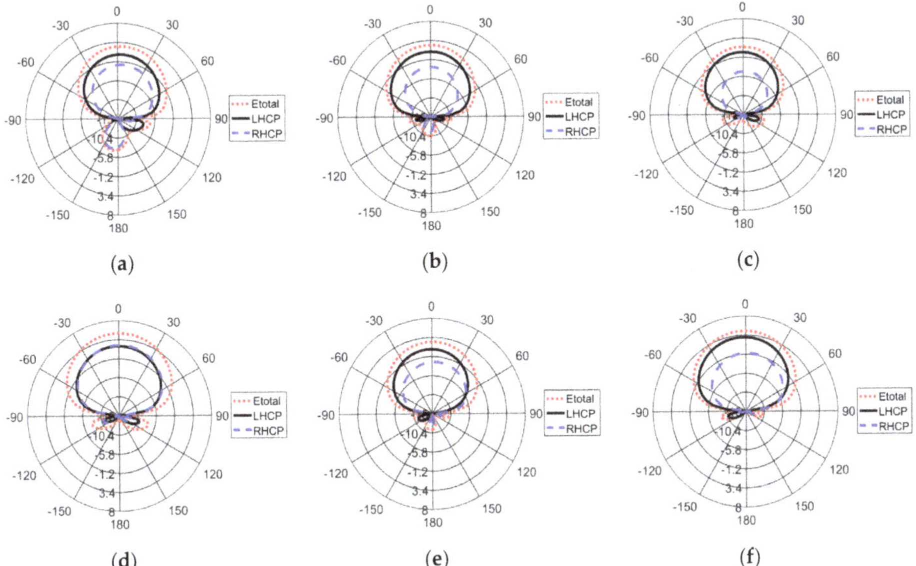

Figure 9. Radiation patterns for center frequencies of each resonance for Case 3 (**a**) f = 2.379 GHz for On-Off, (**b**) f = 2.439 GHz for On-Off, (**c**) f = 2.379 GHz for Off-On, (**d**) f = 2.439 GHz for Off-On, (**e**) f = 2.379 GHz for On-On and (**f**) f = 2.439 GHz for On-On.

In Figures 7–9 there are several different polarizations depending on the port configuration. For example, in Figure 7a–d we have LHCP and RHCP but with equal amplitudes. The result is linear polarization because the LHCP negates the one axis of RHCP and vice versa. On the other hand, in Figure 7e–f, since the LHCP is the same as the total E–field, while the RHCP is significantly weaker, we have circular polarization. In the rest of the patterns (Figures 8 and 9), in some circumstances, neither LHCP nor RHCP are equal with the total E-field. This is indicative of a combination of those circular polarizations resulting to elliptical one. To sum up, by exciting different ports, linear or circular – elliptical polarization can be achieved. In fact the exact circular polarization could be a result of different amplitude between feeding ports because the original polarization is elliptical and so we can make use of Left Hand (LH), Right Hand (RH) circular-elliptical or even linear polarization. This can also be seen in Figure 10 where the axial ratio is observed for the different configurations of ports in Case 2 (D1 On and D2 Off). Each port excitation is in the form A, θ for $A \cdot e^{j\theta}$ (his configuration is selected at random to extract a distinct figure).

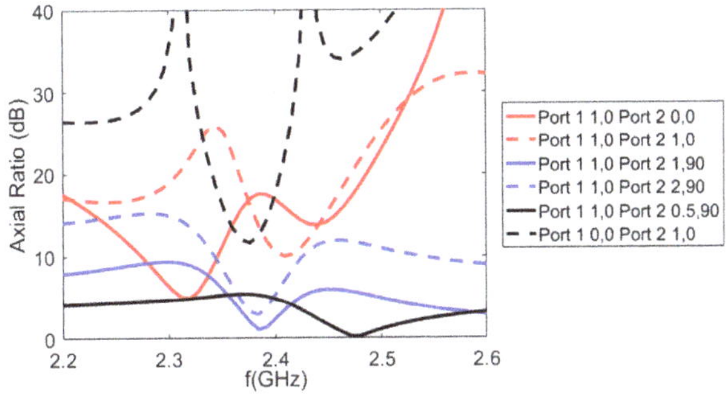

Figure 10. Axial ratio vs. frequency for different excitation signals for Case 2.

Here we can have all the aforementioned polarization by controlling the amplitude and phase of the excitation of each port. So, by not exciting one of the ports, we can utilize linear polarization while by using a phase difference of 90° degrees with the proper amplitude ratio we can utilize circular polarization. The gain along with the efficiency of each case is given in Figure 11, where we observe that the Gain (dBi) of the antenna ranges between 3 and 5.5 dB while the efficiency is above 50% for the operation BW of each case. The efficiency is reduced compared to a common patch but this is a result of the bandwidth enhancement [22] of each resonance and not of the use of PIN diodes.

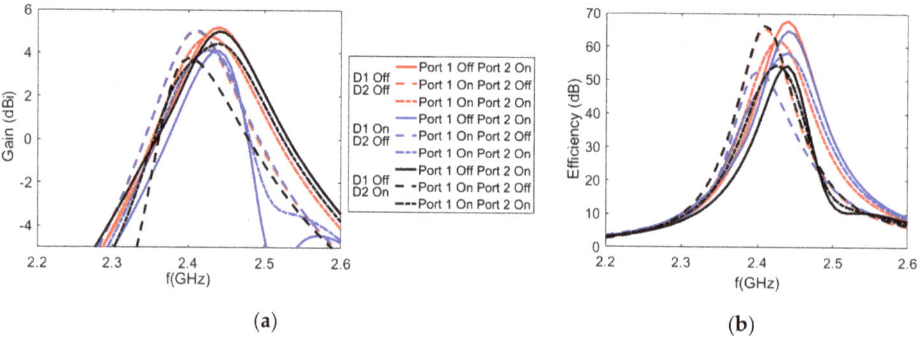

Figure 11. (**a**) Gain vs. frequency and (**b**) Efficiency vs. frequency for each case.

4. Proper use of Case 4 (Switch 1 on–Switch 2 on)

The Case 4 of Table 5 have not been thoroughly explored due to significant coupling (S_{21} being above −10 dB) between the two ports. As an effort to utilize the Case 4 configuration, proper matching is necessary using matching circuits after each port. The matching is not realized through some of the techniques of the literature but through an optimization procedure. The goal of the optimization is to match the port 2 to Z_{in2} = 50 Ohm while keeping the port 1 to Z_{in1} = 50 Ohm. We used Optenni Lab to design it [23]. The optimized matching network for both ports as it was produced by Optenni Lab is depicted in Figure 12. The elements T2, T5 and T7 are transfer lines and the rest of them (T1, T3, T4, T6 and T8) are open circuited stubs. The results for the S-parameters after matching are given graphically in Figure 13, and the values of the important frequencies in Table 8, respectively.

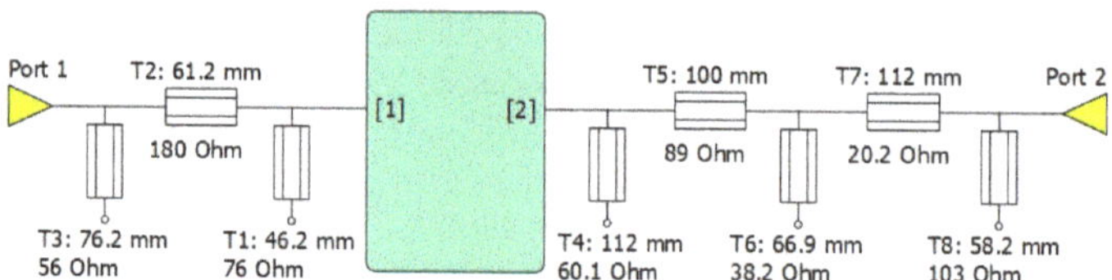

Figure 12. Matching circuits of ports 1 and 2 after optimization.

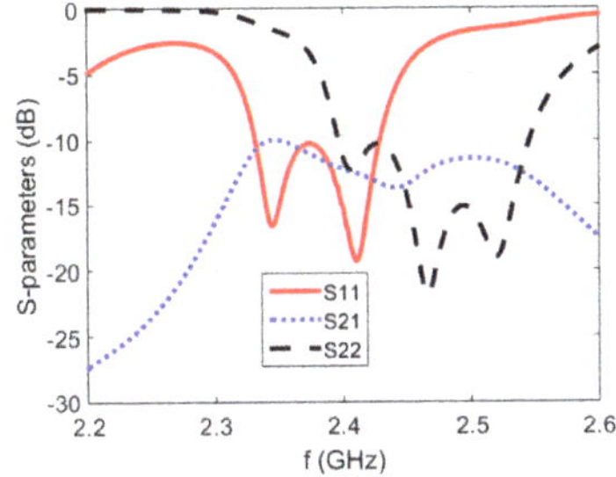

Figure 13. S-parameters of well-matched Case 4 (Switch 1 and 2 On).

Table 8. Important frequencies of the well-matched Case 4.

	Resonance 1 (S_{11})	Resonance 2 (S_{22})
f_L (GHz)	2.3309	2.3979
f_H (GHz)	2.4289	2.5469
f_C (GHz)	2.3799	2.4724
BW (MHz)	98	149
BW (%)	4.11	6.02

5. Comparison with State of the Art Designs from the Literature

A comparison of the proposed antenna characteristics to similar designs from the recent literature is summarized in Table 9. It is obvious that the proposed work is able to produce any kind of polarization except the diagonal one, while its dimensions are kept small enough to be able to be attached to a mobile device.

Table 9. Characteristics Comparison of proposed SPA with other works from the literature.

Ref	Size (mm^3)	Number of Switches	f_c (GHz)	BW (MHz)	FBW (%)	Polarization
[8]	$60 \times 40 \times 1.6$	4	8.4	700	8.3	Linear
[10]	$30 \times 40 \times 0.44$	2	2.75	2500	90.9	Linear
[12]	$19 \times 19 \times 1.6$	2	2.42	95	3.92	Horizontal Vertical
[14]	$72 \times 72 \times 12.5$	4	3.635	790	21.7	Horizontal Vertical
			5.11	1320	25	
[15]	$40 \times 40 \times 3.175$	2	5.9	360	6.1	LP
			5.95	700	11.7	CP
[16]	$120.26 \times 120.26 \times 3.175$	24	2.32	460	19.8	RHCP LHCP
[17]	$50 \times 50 \times 3.048$	-	2.4417	44.5	1.82	Horizontal
			2.4093	40.5	1.68	Vertical
			2.4245	10	0.41	RHCP LHCP Diagonal
Proposed antenna	$50 \times 50 \times 3.048$	2	2.4386	215.2	8.82	Horizontal Vertical RHCP LHCP Elliptical

6. Conclusions

In this paper an elliptical supershape patch radiator is presented. The initial Supershape patch had two resonant frequencies with fractal BW of 1.48%, and 1.11%, respectively. By implementing the proper feeding network, we over doubled the fractal BW of each resonance and by bridging the gap between them a wider single operation band was achieved reaching an 8.8% fractal BW. Also, by using two PIN diodes, one on each feed, the operating bandwidth can be changed, thus creating a frequency reconfigurable antenna. Finally, by using different values for the amplitude and phase of each port, all kinds of polarization can be excited.

Author Contributions: Conceptualization, G.K. and M.C.; methodology, A.K.; software, A.K. and G.X.; validation, A.K. and G.X.; formal analysis, G.K. and M.C.; investigation, A.K.; resources, A.K.; data curation, A.K.; writing—original draft preparation, A.K.; writing—review and editing, A.K. and M.C.; visualization, A.K. and M.C. All authors have read and agreed to the published version of the manuscript.

Funding: This project has received funding from the Hellenic Foundation for Research and Innovation (HFRI) and the General Secretariat for Research and Technology (GSRT), under grant agreement No [2047].

Acknowledgments: The proper matching of the antenna from Z_{in2}=130 Ohm to 50 Ohm as proposed in Section 2 and carried out in Section 4 have been simulated using Optenni Lab.

Conflicts of Interest: The authors declare no conflict of interest.

References

1. Ojaroudi Parchin, N.; Jahanbakhsh Basherlou, H.; Al-Yasir, Y.I.A.; Abd-Alhameed, R.A.; Abdulkhaleq, A.M.; Noras, J.M. Recent Developments of Reconfigurable Antennas for Current and Future Wireless Communication Systems. *Electronics* **2019**, *8*, 128. [CrossRef]
2. Row, J.; Tsai, C. Pattern Reconfigurable Antenna Array with Circular Polarization. *IEEE Trans. Antennas Propag.* **2016**, *64*, 1525–1530. [CrossRef]
3. Bouslama, M.; Traii, M.; Denidni, A.T.; Gharsallah, A. Beam-Switching Antenna with a New Reconfigurable Frequency Selective Surface. *IEEE Antennas Wirel. Propag. Lett.* **2016**, *15*, 1159–1162. [CrossRef]
4. Jin, P.G.; Li, M.; Wang, W.; Li, L.; Liao, S. Planar beam-switching dipole antenna. *IET Microw. Antennas Propag.* **2019**, *13*, 997–1002. [CrossRef]
5. Jin, G.; Li, M.; Liu, D.; Zeng, G. A Simple Planar Pattern-Reconfigurable Antenna Based on Arc Dipoles. *IEEE Antennas Wirel. Propag. Lett.* **2018**, *17*, 1664–1668. [CrossRef]
6. Qin, P.; Wei, F.; Guo, J.Y. A Wideband-to-Narrowband Tunable Antenna Using A Reconfigurable Filter. *IEEE Trans. Antennas Propag.* **2015**, *63*, 2282–2285. [CrossRef]
7. Nguyen-Trong, N.; Piotrowski, A.; Hall, L.; Fumeaux, C. A Frequency- and Polarization-Reconfigurable Circular Cavity Antenna. *IEEE Antennas Wirel. Propag. Lett.* **2017**, *16*, 999–1002. [CrossRef]
8. Khan, T.; Rahman, M.; Akram, A.; Amin, Y.; Tenhunen, H. A Low-Cost CPW-Fed Multiband Frequency Reconfigurable Antenna for Wireless Applications. *Electronics* **2019**, *8*, 900. [CrossRef]
9. Li, T.; Zhai, H.; Wang, X.; Li, L.; Liang, C. Frequency-Reconfigurable Bow-Tie Antenna for Bluetooth, WiMAX, and WLAN Applications. *IEEE Antennas Wirel. Propag. Lett.* **2015**, *14*, 171–174. [CrossRef]
10. Abutarboush, F.H.; Atif Shamim, A. A Reconfigurable Inkjet-Printed Antenna on Paper Substrate for Wireless Applications. *IEEE Antennas Wirel. Propag. Lett.* **2018**, *17*, 1648–1651. [CrossRef]
11. Mathur, R.; Dwari, S. Compact planar reconfigurable UWB-MIMO antenna with on-demand worldwide interoperability for microwave access/wireless local area network rejection. *IET Microw. Antennas Propag.* **2019**, *13*, 1684–1689. [CrossRef]
12. Raman, S.; Mohanan, P.; Timmons, N.; Morrison, J. Microstrip-Fed Pattern- and Polarization- Reconfigurable Compact Truncated Monopole Antenna. *IEEE Antennas Wirel. Propag. Lett.* **2013**, *12*, 710–713. [CrossRef]
13. Shirazi, M.; Li, T.; Huang, J.; Gong, X. A Reconfigurable Dual-Polarization Slot-Ring antenna Element With Wide Bandwidth for Array Applications. *IEEE Trans. Antennas Propag.* **2018**, *66*, 5943–5964. [CrossRef]
14. Nie, Z.; Zhai, H.; Liu, L.; Li, J.; Hu, D.; Shi, J. A Dual-Polarized Frequency-Reconfigurable Low-Profile Antenna with Harmonic Suppression for 5G Application. *IEEE Antennas Wirel. Propag. Lett.* **2019**, *18*, 1228–1232. [CrossRef]

15. Qin, P.; Weily, R.A.; Jay, Y. Polarization Reconfigurable U-Slot Patch Antenna. *IEEE Trans. Antennas Propag.* **2010**, *58*, 3383–3388. [CrossRef]

16. Cai, Y.; Gao, S.; Yin, Y.; Li, W.; Luo, Q. Compact-Size Low-Profile Wideband Circularly Polarized Omni-directional Patch Antenna with Reconfigurable Polarizations. *IEEE Trans. Antennas Propag.* **2016**, *64*, 2016–2021. [CrossRef]

17. Koutinos, G.A.; Ioannopoulos, A.G.; Chryssomallis, T.M.; Kyriacou, A.G. A Wideband Matching Technique for Polarization Versatile Applications. In Proceedings of the 36th PIERS Proceedings, Prague, Czech Republic, 6–9 July 2015; pp. 2081–2085.

18. Paraforou, V.; Tran, D.; Caratelli, D. A novel supershaped slot-loaded printed dipole antenna with broadside radiation for dual-band WLAN applications. In Proceedings of the the the 8th European Conference on Antennas and Propagation (EuCAP 2014), The Hague, The Netherlands, 6–11 April 2014; IEEE: Piscataway, NJ, USA, 2014.

19. Gielis, J. A Generic geometric transformation that unifies a wide range of natural and abstract shapes. *Am. J. Bot.* **2003**, *90*, 333–338. [CrossRef] [PubMed]

20. Koutinos, G.A.; Kyriacou, A.G.; Volakis, L.G.; Chryssomallis, T.M. Bandwidth Enhancement of Antennas Designed by Band-Pass Filter Synthesis Due to Frequency Pulling Techniques. In *Antennas and Propagation*; IET: Stevenage, UK, under Review.

21. Rahola, J. Antenna Matching Circuit Optimization Including Stop Band Definitions. In *IEEE International Symposium on Antennas and Propagation (APSURSI), 3–8 July 2011*; IEEE: Spokane, WA, USA, 2011; pp. 1304–1307.

22. Pues, H.F.; Van de Capelle, A.R. An impedance-matching technique for increasing the bandwidth of microstrip antennas. *IEEE Trans. Antennas Propag.* **1989**, *37*, 1345–1354. [CrossRef]

23. Optenni Lab Antenna Analysis Software. Available online: www.optenni.com (accessed on 10 March 2020).

 *electronics*

Article

Reconfigurable 3-D Slot Antenna Design for 4G and Sub-6G Smartphones with Metallic Casing

Peng Yang

School of Electronic Science and Engineering, University of Electronic Science and Technology of China,
No. 2006 Xiyuan Street, Chengdu 611731, China; yangpeng@uestc.edu.cn

Received: 1 January 2020; Accepted: 24 January 2020; Published: 25 January 2020

Abstract: The design of a reconfigurable three-dimensional (3-D) slot antenna for 4G and sub-6G smartphone application is presented in this paper. The antenna is located at the bottom of the smartphone and integrated with a metallic casing. Positive-Intrinsic-Negative (PIN) diodes are loaded at the dual-open slot and the folded U-shaped slot, respectively, which are used to realize four working states. The antenna has a compact volume of $42 \times 6 \times 6$ mm^3, which can cover the long term evolution (LTE) bands of 698–960 MHz and 1710–2690 MHz, and the sub-6G bands of 3300–3600 MHz & 4800–5000 MHz. The design processes are presented and the structure is optimized, fabricated and measured. The comparison to other state-of-the-art antennas shows that the proposed design has multiband characteristics with small size.

Keywords: reconfigurable antenna; smartphone; slot antenna; sub-6G

1. Introduction

Recently, the wireless communication network environment has become a mixture of different kinds of heterogeneous networks. The WRC-15 (Word Radio Communication Conference 2015) identified the frequency spectrum of 3400–3600 MHz for global mobile broadband services in November 2015 [1]. From November 2017, the frequency spectrums of 3300–3600 MHz and 4800–5000 MHz have been identified for fifth-generation (5G) application by the ministry of industry and information technology of the People's Republic of China [2]. In addition, the mobile antenna is required to be operated at 698–960 MHz and 1710–2690 MHz for fourth-generation (4G) mobile communications [3]. To provide these services for different frequencies and communication protocols, the antenna in smartphones should cover multiple bands. On the other hand, in order to meet the consumers' requirements, the mobile phones are being designed to provide a low profile and a large display with a narrow frame for entertainment. Therefore, designing a multiband/wideband mobile phone antenna with limited space is undoubtedly a technological challenge.

A mobile phone with a metal casing (metal frame and metal back cover) has also become popular since the metal casing not only increases mechanical strength but also enhances the esthetical appearance. For the traditional internal antennas [4], such as the monopole, planar inverted-F antenna (PIFA), etc., the performance of the antenna will be adversely affected by the metal casing [5,6]. In [6], the effects of the metal frame on the performance of the antenna are analyzed, and the influences of the metal frame are reduced by inserting multiple gaps and grounded patches. For the approach proposed in [6], however, the multiple gaps on the metal frame may reduce the mechanical strength of the mobile phone. In [7,8], the effects of the metal frame on the antenna are completely eliminated by using the entire or partial frame as the radiation branch of the antenna. These proposed antennas in [7,8], however, occupy a large clearance size of 70×10 mm^2. In another design, an inverted-F antenna IFA frame antenna was proposed to cover multiple bands [9]. The bandwidth of the upper band was not wide enough to cover the sub-6G application bands of 3300–3600 MHz and 4800–5000 MHz.

Slot antennas are widely adopted as an alternative for mobile antenna designs [10–16]. Two folded slot antennas were introduced in [10,11] to cover dual bands. These antennas cannot apply to the mobile phone with metal casing, and the bandwidth is not sufficient to cover the lower band. In another T-shaped open-ended slot design for laptops with a metallic frame, two separate feed strips were adopted to excite the T-shaped slot antenna [12]. Open-slot antennas are reported for tablet computers [13–15] which have wide operating bands to cover 698–960 MHz, 1710–2690 MHz, and 3400–3600 MHz. However, the larger antenna size and the lack of integration with metal casing make it unsuitable for modern mobile phones. In other designs, the varactor diode and PIN diode were used for reconfiguration, resulting in multiple antenna states [16–18]. This is a very promising design, but only covers part of the high-frequency bandwidth, which limits its application in future 5G mobile phones. In [19], a multi-band loop antenna with six resonant modes is proposed. The antenna can cover the bands from 660 MHz to 5850 MHz. In [20], a reconfigurable loop antenna with two parasitic grounded strips for modern smartphone devices is presented. However, the antenna can only cover 4G bands.

There is no doubt that it is a challenging for antenna engineers to integrate antennas into a metal casing environment and to achieve multi-band/wideband coverage. In this paper, a reconfigurable 3-D slot antenna fed by an L-shaped feed strip is proposed. The 3-D slot antenna can achieve miniaturization and multi-band operation by folding the slot structure and introducing PIN diodes, respectively. The proposed 3-D slot antenna shows four bands that cover 698–960 MHz, 1500–3000 MHz, 3300–3800 MHz, and 4400–5000 MHz for LTE and sub-6G operations.

2. Antenna Configuration

Figure 1a shows the proposed 3-D slot antenna mounted on a mobile phone with a size of $135 \times 75 \times 6$ mm. The mobile phone has a metallic frame and back cover. A slot starts from one narrow side of the frame, (see 'A' in Figure 1b), along 'B' at the back cover, folded to the upper layer 'C', and short-ended at 'D'. Three black rectangles noted as 'd1', 'd2' and 'd3' are PIN diodes. The 3-D slot is fed by an L-shaped strip line, as shown in Figure 1c. The feed line is between the back cover and the system circuit board ground. A 0.8 mm-thick print circuit board type of (FR4) substrate with a relative permittivity of 4.4 and a loss tangent of 0.02 was used as the system circuit board which mounted on the back metal cover. There is no gap between the system circuit board and the metal casing. At the end of the feed line, a lumped port is used to simulate the source.

The topology of the PIN diodes (d1, d2 and d3) is shown in Figure 2. The back-to-back configuration shown in Figure 2a was used here to overcome the problem of equal voltages on both sides of the slot. The PIN diode is Infineon BAR64-02V. The diode is switchable between 'OFF' and 'ON' states depending on a DC bias. Combining the data sheet of the diode and the test results, the equivalent circuit of the diode is shown in Figure 2b. When the diode is 'ON', it has a forward resistance of 2.1 Ω [21]. When the diode is 'OFF', we obtained the values of the corresponding components experimentally at different frequency bands and show them in Table 1.

Table 1. Equivalent circuit parameters for the state 'OFF'.

Frequency (MHz)	R	C	L
800–1000	3933 Ω	148 fF	2.1 nH
1500–3000	3014 Ω	148 fF	5.5 nH
3000–4000	2079 Ω	154 fF	200 pH
4000–5000	2419 Ω	113 fF	181 pH

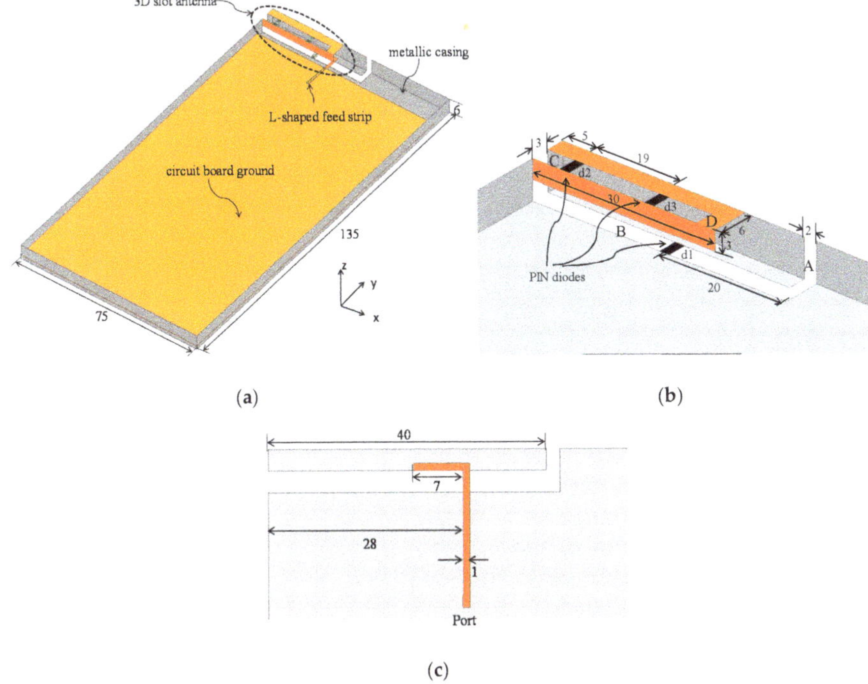

(a)

(b)

(c)

Figure 1. Geometry of the structure. (**a**) Side view of the proposed antenna; (**b**) 3-D view of the antenna; (**c**) the slot and the L-shape feed strip (all in millimeters).

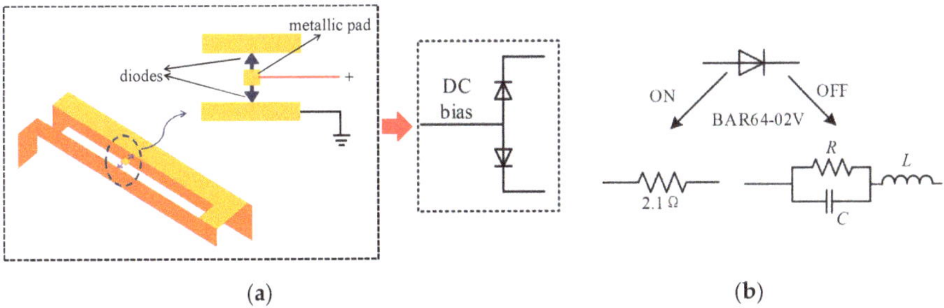

(a)

(b)

Figure 2. The description of the PIN diodes. (**a**) Topology of the back-to-back diodes; (**b**) Equivalent circuit model of BAR64-02VPIN diode in two states.

3. Design Process

3.1. The Principle of the Design

The idea of the design starts from a simple open-ended slot antenna, as shown in Figure 3a. A slot is etched on the edge of a metal. One end of the slot is opened and the other end is shorted. The slot can support a quarter of a standing wave. Details of the radiation mechanism of the slot can be found in [22]. For the lowest frequency (around 700 MHz) that we need, the length of the slot is about 107 mm, which is too long to be used for smart phones. To reduce the length, the slot is modified by three steps: firstly, the open-ended position is moved from the long frame side of the phone to the narrow frame side, as shown in Figure 3b; secondly, the slot length is extended by folding the structure to upper layer, see Figure 3c; and thirdly, an L-shaped feed line was used to feed the slot. By tuning the location and the length of the feed line carefully, four resonate modes can be excited.

It can be seen from Figure 4a that the slot is expected to resonate at a low frequency around 740 MHz owning to 0.25 λ slot mode (since the slot is covered by a FR4 substrate, the actual length is shorter than the length in free space). In addition, a second resonance produced at 2.1 GHz, which is mainly due to a fictitious short circuit near the capacitive feed strip [23,24]. Similarly, two resonances (3.2 GHz & 5 GHz) can also be excited. However, the slot still cannot cover the bands of 824–960 MHz and 1710–2690 MHz due to the narrow bandwidth characteristics.

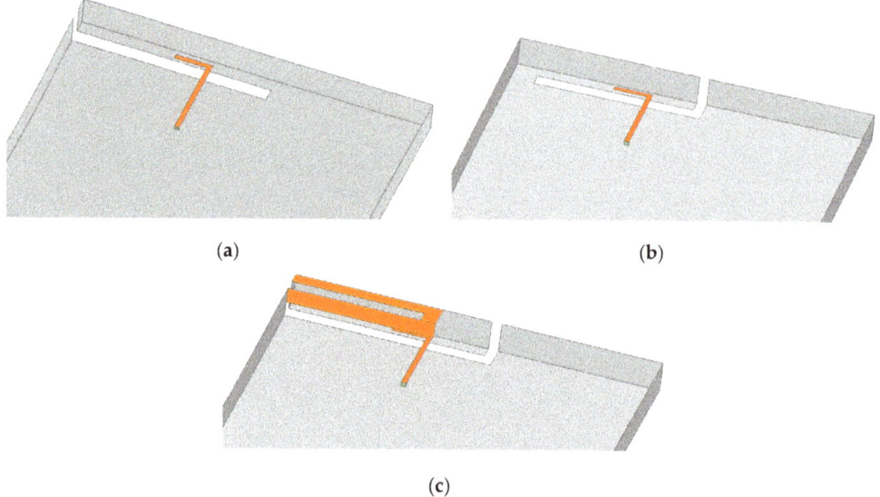

Figure 3. Evolution of the configuration. (**a**) Open-ended slot antenna with the open position at the long frame side; (**b**) open-ended slot antenna with the open position at the narrow frame side; (**c**) folded 3-D slot.

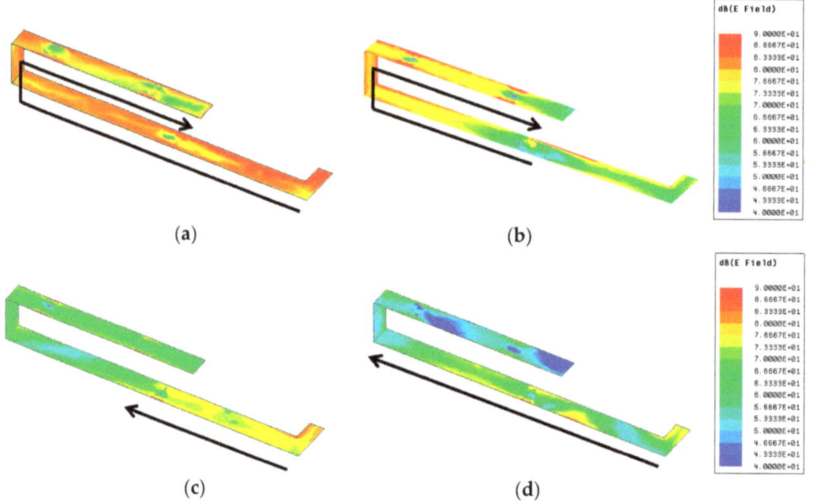

Figure 4. Electric field distribution of state 1 at different resonate modes: (**a**) 0.74 GHz (0.25λ); (**b**) 2.1 GHz (0.5λ); (**c**) 3.2 GHz (0.25λ); (**d**) 5 GHz (0.75λ).

To solve the problem, three PIN diodes are integrated into the slot at proper positions to provide more resonances. On the upper layer, the diodes d2 and d3 are used to change the total length of the slot, which will generate two additional resonances around 820 MHz and 920 MHz. The diodes d1 on the bottom layer are used to generate the LTE band from 1710–2690 MHz. These diodes make the antenna reconfigurable to enable multi-band operation. The states of these diodes are summarized in Table 2. The simulated reflection coefficients of the antenna for different states are shown in Figure 5.

Table 2. States of three back-to-back diodes.

	d1	d2	d3	Frequency
State 1	OFF	OFF	OFF	700~770 MHz 2000~3600 MHz 4800~5200 MHz
State 2	OFF	OFF	ON	770~870 MHz
State 3	OFF	ON	OFF	850~960 MHz
State 4	ON	OFF	OFF	1300~3000 MHz

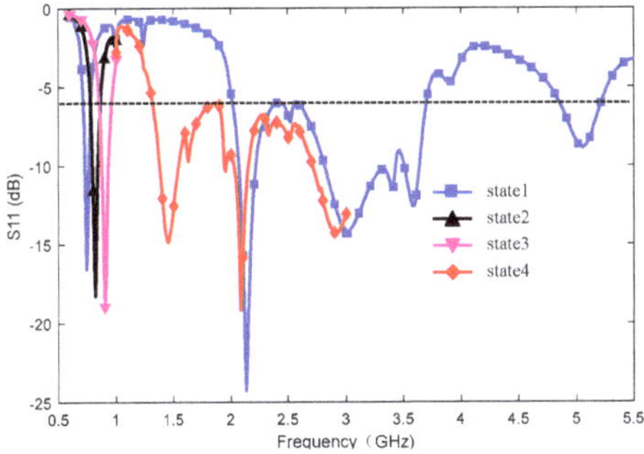

Figure 5. Simulated reflect coefficients of the proposed antenna for different diode states.

3.2. Key Parameters of the Antenna

The structure of the design is very simple and only few parameters need to be determined. The lower resonate frequencies depend on the total length of the 3-D slot and the locations of diodes d2 and d3, while the impedance between 1710–2690 MHz is mainly affected by the locations of diodes d1 and the feed line.

To optimize the antenna performance of 1710–2690 MHz, the locations of the back-to-back diodes d1 and the feed line, and their effects are studied. In the simulations, the switching effect of the diodes is equivalent to the circuit shown in Figure 2b. Since the resonate frequency mainly depends on the location of diodes d1, we can first estimate the value of $p1$. Choosing the center frequency at 2200 MHz, the slot length from point 'A' in Figure 1b to d1 is about quarter wavelength of 2200 MHz. The length of the slot should be shorter than this because it is printed on a FR4 board with a permittivity of 4.4. Hence, the estimated value of $p1$ is around 20 mm. The S11 of the antenna working at state 4 for different $p1$ and $p2$ are studied and the results are shown in Figure 6b,c. In Figure 6b, the parameter $p1$ is fixed at 19 mm while in Figure 6c the parameter $p2$ is fixed at 28 mm. It can be seen that when $p1 = 19$ mm and $p2 = 28$ mm, the antenna can match well in this frequency band.

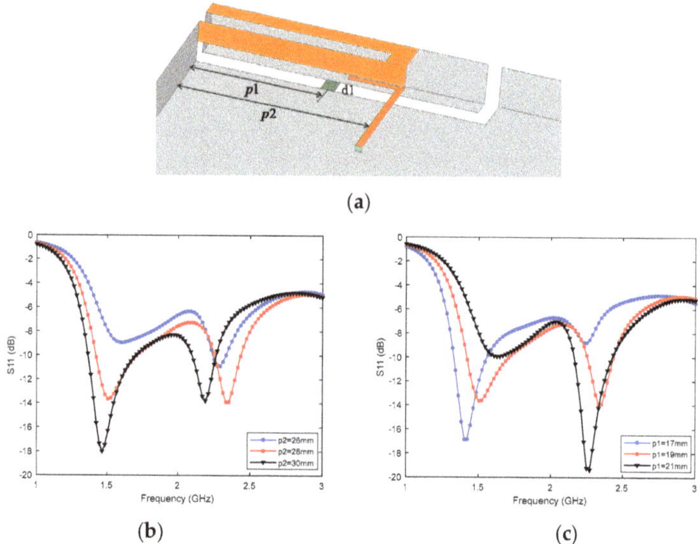

Figure 6. Parameters studies of diode d1 and feed line. (**a**) Locations of diode d1 and the feed line. (**b**) and (**c**) S11 performance for different $p1$ and $p2$.

4. Measurement Results

The proposed reconfigurable 3-D slot antenna was fabricated and tested. Figure 7a,b show the front and back of the prototype, respectively. The detail of the control circuit of the diodes is shown in Figure 7c. When the positive pole of the voltage (red line) is connected to the center of the back-to-back diodes, as shown in Figure 2a, and the negative pole (black line) is connected to ground, the back-to-back diodes are 'ON'. When the red line is removed, the diodes are 'OFF'.

Figure 8 shows the simulated and measured reflection coefficient of the three working states of the prototype. In this figure, good agreement between the simulated results and the measured data was obtained for overall bands. The measured −6 dBi impedance bandwidth can cover 690–1000 MHz, 1500–3800 MHz and 4400–5000 MHz, which can satisfy the desired bands for WWAN/LTE and 5G operations, respectively. The measured total efficiency and gain for different states of the proposed antenna are shown in Figure 9. For the lower band of 800–1000 MHz, the measured efficiency is not high enough. There are three reasons: The main reason is the small size of the antenna. The volume of the antenna is only $42 \times 6 \times 6$ mm^3 and its electric size is only 0.11λ around 800 MHz, so the radiation efficiency is relatively low. Secondly, the low efficiency is due to the resistor in the diode. The diode is not an ideal conductor but has a 2.1 Ohm resistor even in the 'ON' state. Hence, the back-to-back diodes have a 4.2 Ohm resistor, which will cause a big loss. Thirdly, the folded slot is printed on a FR4 board with a loss tangent of 0.02, which will further reduce the efficiency. For the higher bands of 1500–3000 MHz, 3100–3800 MHz and 4400–5000 MHz, the measured total efficiencies are between 40–80%. Similarly, the gain of the antenna varies from −5 dB to 5 dBi throughout the desired bands. The measured far field gain patterns at 800, 2100, 3500, and 4900 MHz are shown in Figure 10a-d. These patterns are not symmetrical due to the asymmetrical antenna structure.

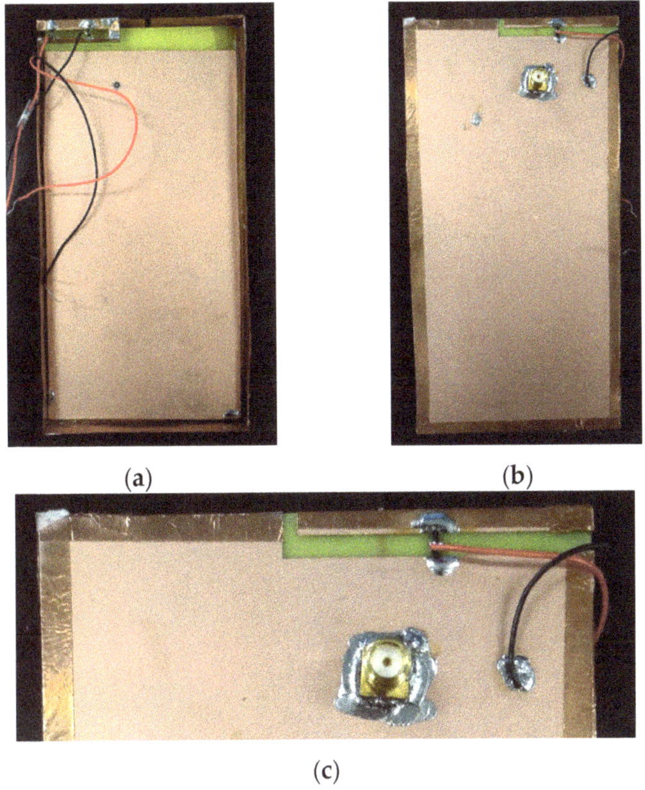

(a)   (b)

(c)

Figure 7. Prototype of the antenna. (**a**) Front; (**b**) back; (**c**) details of the control circuit of the diode.

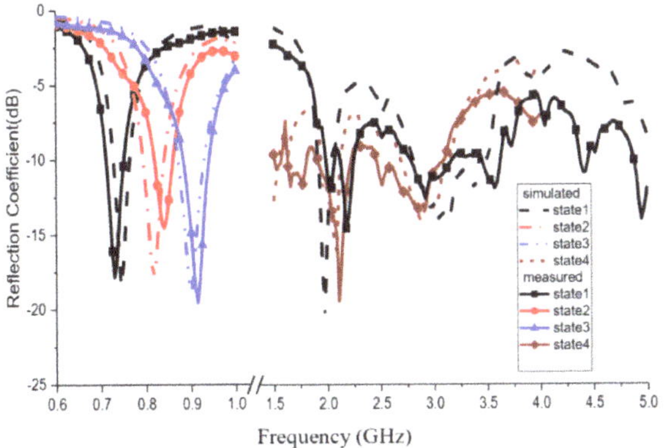

Figure 8. Measured reflection coefficient of different states.

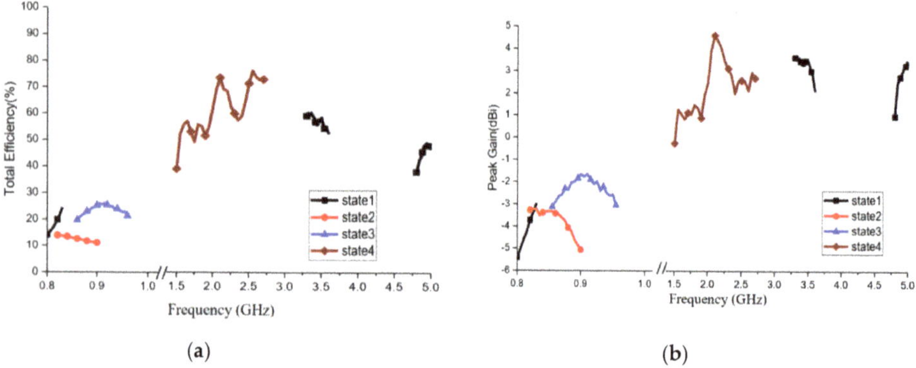

(a)   (b)

Figure 9. Measured efficiency (**a**) and gain (**b**).

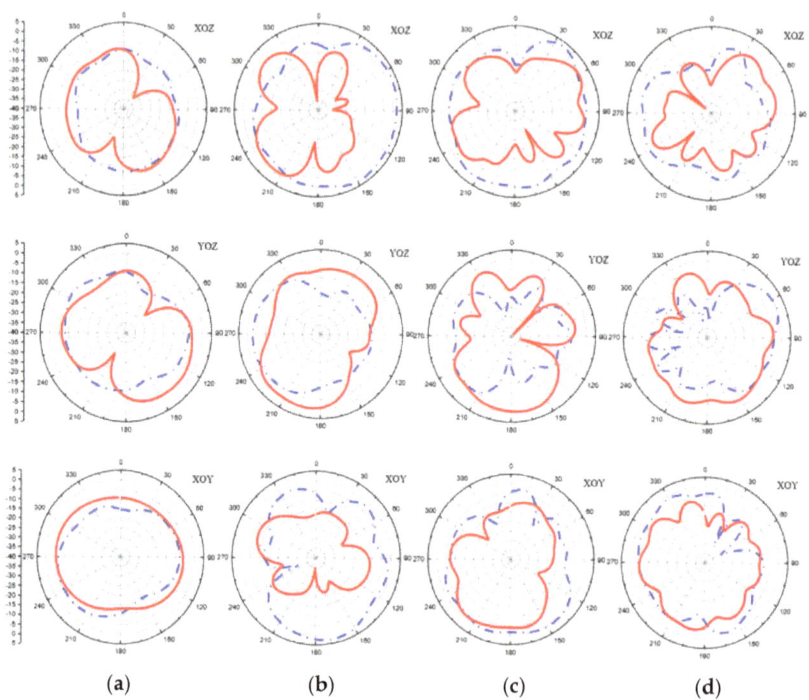

(a)   (b)   (c)   (d)

Figure 10. Measured 2-D radiation patterns of the proposed antenna at three principal planes (red solid line E_θ, blue dashed line E_φ, unit: dB): (**a**) 800 MHz; (**b**)2100 MHz; (**c**) 3500 MHz; (**d**) 4900 MHz.

5. State-of-Art Comparison

In order to evaluate the performance of the proposed reconfigurable 3-D slot antenna, its performance is compared with recently published designs. The results are summarized in Table 3. As can be seen, the proposed antenna exhibits a smaller clearance zone on the premise of integrating with the metal casing (metal frame and metal back cover). Also, it has more bands than most of the designs in the literatures.

Table 3. Comparison of proposed antenna performance with recent antenna designs.

Ref.	MF [1]	MBC [2]	Clearance Zone (mm²)	BWlow [3] (MHz)/η/Gain	BWhigh [4] (MHz)/η/Gain
This work	Yes	Yes	42 × 6	698~1000/10~28%/−5~−1 dBi	1500~3000/40~75%/0~5 dBi 3100~3800/50~60%/2~4 dBi 4400~5000/40~50%/0~3 dBi
[7]	Yes	No	70 × 10 & 70 × 5	824~960/60~79%/1~2 dBi	1710~2690/54~79%/1~4 dBi
[8]	Yes	No	70 × 10	801~1002/>42%/0~3 dBi	1695~3000/>51%/2~5 dBi
[16]	Yes	Yes	68 × 11	698~960/45~90%/N.A.	1710~2690/50~95%/N.A.
[18]	No	No	75 × 10	820~960/40~90%/N.A.	1710~2690/40~80%/N.A. 3400~3600/40~70%/3~6 dBi
[19]	Yes	Yes	70 × 10	660~1100/25~65%/N.A.	1710~3020/60~90%/N.A. 3370~3900/50~90%/N.A. 5150–5850(70~90%)/N.A.
[20]	Yes	No	67 × 10	824~960/20~60%/−2~2 dBi	1710~2690/50~80%/2~4 dBi

[1] MF = Metal Frame. [2] MBC = Metal Back Cover. [3] BWlow = Bandwidth at lower-band. [4] BWhigh = Bandwidth at higher-band. η = efficiency. N.A. = Not Applicable.

6. Conclusions

In this paper, a novel and compact reconfigurable 3-D slot antenna with a metal casing which can be applied to 5G smartphones is proposed and studied. Three back-to-back PIN diodes are used to achieve four working states with the reflection coefficient less than −6 dB throughout the 698–1000 MHz and 1500–3000 MHz bands and the sub-6G bands (3100–3800 MHz & 4400–5000 MHz). The proposed antenna has a very small ground clearance. It is a promising candidate for LTE/5G smartphone application.

Author Contributions: Conceptualization, P.Y.; project administration, writing and editing. All authors have read and agreed to the published version of the manuscript.

Funding: This work was supported in part by the National Natural Science Foundation of China under Grant 61871101, and in part by the National Key Research and Development Program of China under Grant 2016YFC0303501.

Conflicts of Interest: The author declares no conflict of interest.

References

1. World Radio Communication Conference Allocates Spectrum for Future Innovation. Available online: http://www.itu.int/net/pressoffice/press_releases/2015/56.aspx (accessed on 23 November 2019).
2. The Ministry of Industry and Information Technology Has Been Informed of the Use of 3300–3600MHz and 4800–5000MHz Frequency Bands in the Fifth Generation Mobile Communication System. Available online: http://www.miit.gov.cn/ (accessed on 23 November 2019).
3. LTE Frequency Bands and Spectrum Allocations. Available online: http://www.radio-electronics.com (accessed on 23 November 2019).
4. Wong, K.L. *Planar Antennas for Wireless Communications*; Wiley: New York, NY, USA, 2003.
5. Hossa, R.; Byndas, A.; Bialkowski, M.E. Improvement of compact terminal antenna performance by incorporating open-end slots in ground plane. *IEEE Microw. Wirel. Compon. Lett.* **2004**, *14*, 283–285. [CrossRef]
6. Guo, Q.; Mittra, R.; Lei, F.; Li, Z.; Ju, J.; Byun, J. Interaction between internal antenna and external antenna of mobile phone and hand effect. *IEEE Trans. Antennas Propag.* **2013**, *61*, 862–870. [CrossRef]
7. Ban, Y.-L.; Qiang, Y.-F.; Chen, Z.; Kang, K.; Guo, J.-H. A dual-loop antenna design for hepta-band WWAN/LTE metal rimmed smartphone applications. *IEEE Trans. Antennas Propag.* **2015**, *63*, 48–58. [CrossRef]
8. Zhang, L.-W.; Ban, Y.-L.; Sim, C.-Y.-D.; Guo, J.; Yu, Z.-F. Parallel dual-loop antenna for WWAN/LTE metal-rimmed smartphone. *IEEE Trans. Antennas Propag.* **2018**, *66*, 1217–1226. [CrossRef]

9. Wong, K.-L.; Wu, Y.-C. Small-size dual-wideband IFA frame antenna closely integrated with metal casing of the LTE smartphone and having decreased user's hand effects. *Microw. Opt. Technol. Lett.* **2016**, *58*, 2853–2858. [CrossRef]

10. Song, Y.; Modro, J.; Wu, Z.; O'Riordan, P. Miniature multiband and wideband 3-D slot loop antenna for mobile terminals. *IEEE Antennas Wirel. Propag. Lett.* **2006**, *5*, 148–151. [CrossRef]

11. Chang, C.-H.; Wong, K.-L. Internal multiband surface-mount monopole slot chip antenna for mobile phone application. *Microw. Opt. Technol. Lett.* **2008**, *50*, 1273–1279. [CrossRef]

12. Wong, K.-L.; Tsai, C.-Y. Low-Profile dual-wideband inverted-T open slot antenna for the LTE/WWAN tablet computer with a metallic frame. *IEEE Trans. Antennas Propag.* **2015**, *63*, 2879–2886. [CrossRef]

13. Wong, K.-L.; Tsai, C.-Y. Dual-wideband U-shape open-slot antenna for the lte metal-framed tablet computer. *Microw. Opt. Technol. Lett.* **2015**, *57*, 2677–2683. [CrossRef]

14. Wong, K.-L.; Huang, C.-Y. Triple-wideband open-slot antenna for the LTE metal-framed tablet device. *IEEE Trans. Antennas Propag.* **2015**, *63*, 5966–5971. [CrossRef]

15. Wong, K.-L.; Li, Y.-J. Low-profile open-slot antenna with three branch slots for triple-wideband LTE operation in the metal-framed smartphone. *Microw. Opt. Technol. Lett.* **2015**, *57*, 2231–2238. [CrossRef]

16. Stanley, M.; Huang, Y.; Wang, H.; Zhou, H.; Tian, Z.; Xu, Q. A Novel reconfigurable metal rim integrated open slot antenna for octa-band smartphone applications. *IEEE Trans. Antennas Propag.* **2017**, *65*, 3352–3363. [CrossRef]

17. Ban, Y.-L.; Qiang, Y.-F.; Wu, G.; Wang, H.; Wong, K.-L. Reconfigurable narrow-frame antenna for LTE/WWAN metal-rimmed smartphone applications. *IET Microwaves Antennas Propag.* **2016**, *10*, 1092–1100. [CrossRef]

18. Chen, Q.; Lin, H.; Wang, J.; Ge, L.; Li, Y.; Pei, T.; Sim, C.-Y.-D. Single ring slot-based antennas for metal-rimmed 4G/5G smartphones. *IEEE Trans. Antennas Propag.* **2019**, *67*, 1476–1487. [CrossRef]

19. Xu, H.; Wang, H.; Gao, S.; Zhou, H.; Huang, Y.; Xu, Q.; Cheng, Y.J. A compact and low-profile loop antenna with six resonant modes for LTE Smartphone. *IEEE Trans. Antennas Propag.* **2016**, *64*, 3743–3751. [CrossRef]

20. Zhang, H.-B.; Ban, Y.-L.; Qiang, Y.-F.; Guo, J.; Yu, Z.-F. Reconfigurable loop antenna with two parasitic grounded strips for WWAN/LTE unbroken-metal-rimmed smartphones. *IEEE Access* **2017**, *5*, 4853–4858. [CrossRef]

21. Yang, Z.-X.; Yang, H.-C.; Hong, J.-S.; Li, Y. Bandwidth enhancement of a polarization-reconfigurable patch antenna with stair-slots on the ground. *IEEE Antennas Wirel. Propag. Lett.* **2014**, *13*, 579–582. [CrossRef]

22. Zhang, Z.J. *Antenna Design for Mobile Devices*; John Wiley & Sons Singapore Pte. Ltd.: Singapore, 2017.

23. Behdad, N.; Sarabandi, K. A wide-band slot antenna design employing a fictitious short circuit concept. *IEEE Trans. Antennas Propag.* **2005**, *53*, 475–482. [CrossRef]

24. Behdad; Sarabandi. A multiresonant single-element wideband slot antenna. *IEEE Antennas Wirel. Propag. Lett.* **2004**, *3*, 5–8. [CrossRef]

 *electronics*

Article

A Frequency and Radiation Pattern Combo-Reconfigurable Novel Antenna for 5G Applications and Beyond

Muhammad Kamran Shereen [1,2], **Muhammad Irfan Khattak** [1,*] **and Mu'ath Al-Hasan** [3]

[1] Microwave and Antenna Research Group, Electrical Engineering Department,
   University of Engineering and Technology Peshawar, Peshawar 25120, Pakistan;
   engrkamran@uetpeshawar.edu.pk
[2] United States Pakistan Centre for Advanced Studies in Energy (USPCASE),
   University of Engineering and Technology Peshawar, Peshawar 25120, Pakistan
[3] College of Engineering, Al-Ain University UAE, Abu Dhabi 64141, UAE; muath.alhasan@aau.ac.ae
* Correspondence: M.I.Khattak@uetpeshawar.edu.pk

Received: 12 July 2020; Accepted: 6 August 2020; Published: 25 August 2020

Abstract: This paper presents a novel combo-reconfigurable architecture for the frequency and radiation patterning of a novel antenna system for future fifth-generation (5G) millimeter-wave mobile communication. The tuning system independently controls the frequency and radiation pattern shifts, without letting them affect each other. The proposed antenna consists of two patches, radiating at 28 GHz and 38 GHz. A negative-channel metal–oxide–semiconductor (NMOS) transistor was used as a switch for ON/OFF states. Frequency reconfiguration was controlled by switches SD1 and SD2, while pattern reconfigurability was achieved by SD3–SD18. The desired resonant frequencies of 28 GHz and 38 GHz were achieved by varying patch dimensions through the ON and OFF states of the SD1 and SD2 switches. Similarly, parasitic stubs on the ground are used to control surface currents, which results in pattern reconfiguration. The results were analyzed for 18 different combinations of the switch states. Adding/removing parasitic stubs and switches changed the beam steering angle (by 45° shift) from 0° to 180°, which modified the stub dimensions and changed the beam-width of the main lobe.

Keywords: combo-reconfigurable; frequency and radiation pattern; 5G; millimeter waves; mobile communication

---

1. Introduction

The next generation of telecommunication networks (fifth-generation or 5G) will hit the market by the end of 2021, and will proceed to grow globally. In addition to speed enhancement, 5G is expected to generate a vast IoT (internet of things) ecosystem in which networks can meet the communication needs and expectations of billions of connected devices, with the required standards of speed, latency and cost [1]. 5G mobile communication systems will have a significant impact on modern technology by transmitting information via millimeter waves (mmWaves), and will be quicker, easier to manage, and more reliable than present cellular technology for technological reasons [1]. The frequency band planned for future 5G infrastructure and for meeting bandwidth requirements is expected to be 28 GHz/38 GHz for mobile [2] and 60 GHz/73 GHz for wireless communication [3]. Antennas are a strict elementary part in any wireless system scheme. A lot of research has been concerned with multiband antenna systems operating in millimeter-wave frequency bands, and has been extensively explored [4]. Numerous designs using multi-band methods have been adopted for the 5G millimeter-wave bands. In [5], a circular microstrip antenna with an elliptical slot at 28 GHz and 45 GHz was introduced.

The proposed structure was investigated for configuring a specific absorption rate (SAR) for a unit cell (with an SAR value for 28 GHz of 1.25 W/kg, and 1.38 W/kg for 45 GHz), 1 × 4 array (SAR value at 28 GHz is 1.19 W/kg, at 34 GHz is 1.16 W/kg and at 45 GHz is 1.37 W/kg), and a 5G wearable antenna application. In [6], a multiple input multiple output (MIMO) antenna had three bands for ultra-wide band (UWB) functions, with one band notching an adjusted 5 GHz to stop the wireless local area network (WLAN) frequency. The design of a two-dimensional (2D) slot antenna worked only at 28 GHz, which contained eight essential elements in a 2 × 4 shape [7]. The cellular handset printed circuit board, (PCB) with four single band elements of a microstrip antenna on a low-cost substrate of FR-4 at 28 GHz, was used in [8]. An eight-element microstrip antenna array functioning at 28 GHz and 38 GHz for the next generation of 5G communications was previously investigated [9]. Reference [10] presented a massive MIMO for a 28 GHz and a 38 GHz antenna. In [11], the designs of eight dual-band MIMO antennas for upcoming smartphones were offered. The antenna had a size of 7 × 15.5 mm^2 for 3.3–3.6 GHz and 4.8–5.0 GHz 5G bands. Reference [12] introduced an antenna consisting of a radiator with a triangular shape and exponentially declined edges. The monopole antenna had a condensed area of 10 × 12 mm^2. A 2 × 2 MIMO antenna was shown, with an envelope correlation coefficient lower than 0.001. For the first time, as far as we know, this article has accommodated various approaches of mmWave 5G antennas for cellular smartphones [13].

The published literature [4–13] has reviewed and used variously shaped smart antennas and MIMO antenna systems with radiating patches, and every category has been associated with its own merits and shortcomings for a particular application. Although smart antennas are a powerful concept, the cost associated with the multiple radio frequency (RF) chains and additional digital signal processing (DSP) (an external beamforming network) resources can limit their use in many practical applications. Compared with conventional antennas, the use of reconfigurable antennas is another fascinating and exciting approach for wireless communications network technologies [14]. Moreover, a reconfigurable antenna provides cost-saving functionality, as well as weight, volume and maintenance/repair benefits [14]. This type of antenna helps to reconfigure not only the frequency bandwidth, but also the pattern of radiation and polarization. Reconfigurable antenna provides versatile architectures and agile frequency, using defined operating systems and cognitive radios to deal with multi-service, multi-standard, and multi-band operations, which can be expanded and reconfigured. To implement a dynamical response, they employ different mechanisms such as PIN diodes, varactors, radio-frequency microelectromechanical systems (RF-MEMS), field effect transistors (FETs), parasitic pixel layers, photoconductive elements, mechanical actuators, metamaterials, ferrites and liquid crystals. These mechanisms enable the intentional distribution of current over the antenna's surface, facilitating the reversible modification of their properties.

Much of the literature has focused on tuning a single antenna property, rather than multiple properties. The effective length of the antenna and its operating frequency can be changed by adding or removing part of the antenna's length through electronic, optical, mechanical, or other, means. For instance, an author changed the effective length of a monopole antenna using optical switches, which helped eliminate some of the switch and bias line effects that can occur with other kinds of switches [15]. A similar design approach was taken in [16]. In this case, a balanced dipole fabricated on high-resistivity silicon was equipped with two silicon photo-conducting switches. Using four PIN diodes, a researcher developed a reconfigurable printed dipole antenna to deliver three operating bands between 5.2 GHz and 5.8 GHz [17]. Others have applied the same approach to microstrip patches [18], microstrip dipoles [19] and Yagi antennas [20]. Radiating structures based on fractal shapes that share the same underlying principle have been studied by numerous researchers [21,22]. With the reflective surface physically removed and isolated from the primary feed, reflector antennas are a natural choice for applications that require radiation pattern reconfiguration independent of frequency. Reference [23] demonstrated an example of a radiation-reconfigurable reflector antenna by actively changing the structure of a mesh reflector. In its first embodiment, the reflector contour was changed manually in certain regions, which resulted in changes in beam shape and direction. Later, computer-controlled stepper motors were implemented to pull cables attached to specific points on the reflector mesh, to support automatic pattern reconfiguration [24]. More recently, a similar system for satellite applications was developed to expand capabilities endowed by changes in the system's sub reflector, rather than its main reflector [25]. It is extremely challenging to separate an antenna's frequency characteristics from its radiation characteristics. Indeed, this ability to independently select operating frequency, bandwidth and radiation pattern characteristics is the ultimate goal of reconfigurable antenna designers. Recently, several research groups achieved this kind of reconfigurability, here termed compound or hybrid reconfigurability. A few of the antennas previously discussed are already capable of separating, and of selectable impedance and radiation performance [26,27]. In [26], an annular slot antenna was used as both a frequency- and pattern-reconfigurable device. In [27], a resonant single-turn microstrip spiral antenna provided a broadside radiation operation at two different frequencies, and end-fire radiation characteristics at one of these frequencies. A Yagi-based approach, using reconfigurable slot-loaded parasitic elements as directors and reflectors, was presented in a prior work, supporting tilted or broadside beams at two separate frequencies [28]. In [29], a reconfigurable stacked microstrip antenna delivers a broadside circularly polarized beam at one frequency, and a dipole pattern at a lower frequency, for joint satellite and terrestrial operation.

Aiming at the challenge of controlling the resonant frequencies and direction of the main lobe at the particular angle of interest, this paper provides a suitable solution for rapid development in communication system configurations where the frequency and pattern of the system can be independently tuned within a single antenna. Our research aimed to provide a novel antenna system that can achieve frequency as well as pattern reconfigurability by changing the dimensions of the patch and stubs, respectively. The frequencies targeted here are 28 GHz and 38 GHz (as per Federal Communications Commission (FCC, Washington, DC, USA) standards/guidelines for 5G mobile communication technology [30]), with 360° of rotation in the main beam of the pattern for both the desired frequencies. Until now, no such reconfigurable structure has been cited in mmWave applications.

2. Proposed Antenna Design

The geometric configuration of the proposed antenna is shown in Figure 1. The patch shows a vertically polarized antenna with respect to the feed line, which can be changed to horizontal polarization by varying its orientation. The proposed architecture was selected according to the standard dimensions of 5G cellular phablets. The circuit board was fabricated using a procedure of optic-radiation of photolithography, which processed the photoresist layer to paste the mask onto silicon sheets. The proposed antenna was designed in an Ansys HFSS (High Frequency Structure Simulator) 19.2 3D electromagnetic field simulator for the radio frequency and wireless design, and was measured using a ZVA 40 GHz vector network analyzer. A parametric analysis for the right choice was performed by introducing slots in the conducting patches of both 28 GHz and 38 GHz. To get multiple resonances, the analysis of gain and return loss was performed in accordance with the selection of the optimal model for the final results. As can be seen in Figure 2a,b, the return loss S11 (dB) and gain of different models led us to select model 1, which had a maximum S11 (dB) value (−39.2 dB for 28 GHz, an −38.9 dB for 38 GHz) and reasonable gains. The antenna designed for the nominated frequencies was a cheap and low-profile novel structure. The overall geometry consisted of two targeted frequencies—i.e., partial ground with connected stubs—and a substrate Rogers (Rogers Corporation, Chandler, AZ, USA) Duroid RT 5880, with dimensions of $L \times W \times h = 112 \times 52 \times 0.508$ mm^3 (dielectric = 2.2, loss tangent), where h denotes substrate thickness. The rest of the parameters, as shown in Table 1, were initially calculated via basic antenna equations [31] with respect to the selected resonant frequencies, i.e., 28 GHz and 38 GHz. The microstrip transmission line feeding technique was used to excite the antenna element (thickness of $t = 0.035$ mm).

In Figure 2, it can be seen that the return loss of model 1 is optimal as compared to the other models. This way of choosing optimal return loss is also a type of frequency reconfiguration.

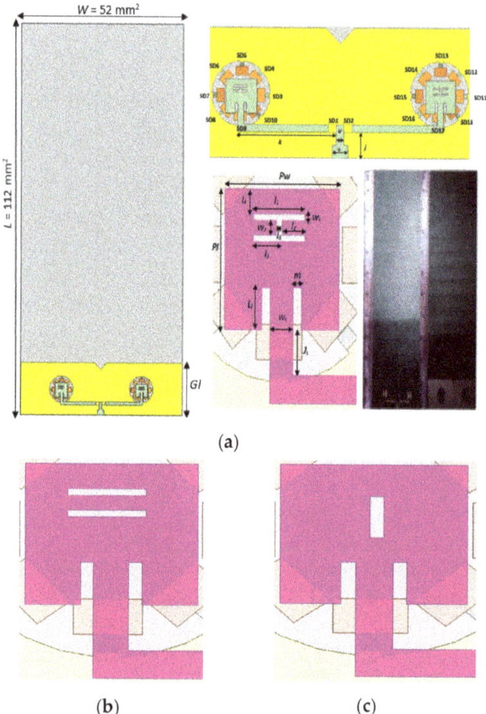

Figure 1. Proposed antenna (**a**) model 1; (**b**) patch of model 2; (**c**) patch of model 3.

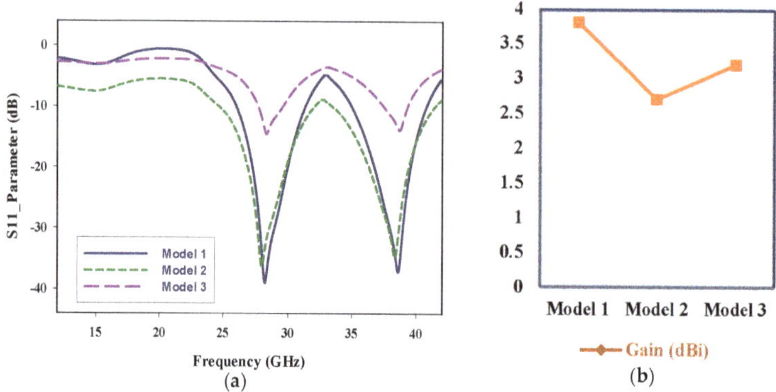

Figure 2. Investigation for parametric analysis. (**a**) Return loss of different models (State 1 of Table 2); (**b**) gain investigation for parametric analysis of different models.

In total, 10 negative-channel metal–oxide–semiconductor (NMOS) transistors were mounted on patched and partial ground. These NMOS transistors were used as RC (Resistor-Capacitor) equivalent circuits, as shown in Figure 3a, with a resistor (*R*) and capacitor (*C*) for ON/OFF states [32]. In order to validate the simulations, the proposed hybrid reconfigurable antenna was fabricated, manufactured or designed for ON/OFF states using PSPICE Software, as shown in Figure 3b.

Table 1. Parameters and their dimensions.

Parameter	Dimension (mm)	Parameter	Dimension (mm)
j	3	$l_{3/28}$	0.3
k	12.5	$l_{3/38}$	0.4
s	2	l_5	0.2
u	1.8	l_f	1.2
l_4	0.2	w_t	0.8
l_1	0.8	Jt_{28}	5.5
$l_{2/28}$	0.3	Jt_{38}	5.2
$l_{2/38}$	0.2	m	0.3
w_1	1.5	Pw_{28}	4
w_2	0.035	Pw_{38}	3.8
Pl_{28}	4	Gd	2
Pl_{38}	3.8	$St_{w/28,38}$	1.9
Gl	15	$St_{l/28,38}$	1

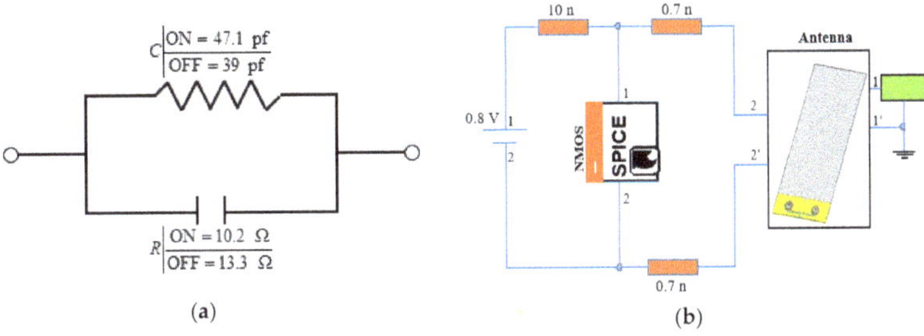

Figure 3. (**a**) Equivalent circuit for ON/OFF states in a negative-channel metal–oxide–semiconductor (NMOS) transistor; (**b**) circuitry for NMOS switch co-simulation for mode 1 (Table 3) of the proposed antenna.

While designing hybrid frequency and a radiation pattern-based antenna, the frequency response, with its related parameters, was considered, such as return loss (RL), voltage standing wave ratio (*VSWR*), reflection coefficient and a dominant TM (Transverse-Magnetic) mode based on cavity theory (Equations (1) and (2) [33]).

$$\Gamma = \frac{Z_{in}(\omega) - Z_0}{Z_{in}(\omega) + Z_0} \tag{1}$$

$$VSWR = \frac{V_{max}}{V_{min}} = \frac{1 + |\Gamma|}{1 - |\Gamma|} \tag{2}$$

The stub radius (R_S) with a fringing effect at frequencies of 28 GHz and 38 GHz was previously identified [34]:

$$R_s = \frac{F}{\left\{1 + \frac{2h}{\pi \varepsilon_r F}\left[\ln\left(\frac{\pi F}{2h}\right) + 1.7726\right]\right\}^{\frac{1}{2}}} \tag{3}$$

where

$$F = \frac{8.791 \times 10^9}{f_0 \sqrt{\varepsilon_r}} \tag{4}$$

Based upon Equations (1)–(4), we observed the different operation modes and states of the antenna, as shown in Tables 2 and 3.

Table 2. State simulation statuses and directions.

State	SD1	SD2	All Other Switches	Direction
State 1	ON	ON	OFF	Pattern is directed in the middle
State 2	ON	OFF	OFF	Uni-directional
State 3	OFF	ON	OFF	Uni-directional

Table 3. Different operational modes for pattern reconfiguration.

Mode	SD1	SD2	SD3	SD4	SD5	SD6	SD7	SD8	SD9	SD10	SD11	SD12	SD13	SD14	SD15	SD16	SD17	SD18
1	ON	OFF	ON	OFF	OFF	OFF	OFF	OFF	OFF	OFF	OFF	OFF	OFF	OFF	OFF	OFF	OFF	OFF
2	ON	OFF	OFF	ON	OFF	OFF	OFF	OFF	OFF	OFF	OFF	OFF	OFF	OFF	OFF	OFF	OFF	OFF
3	ON	OFF	OFF	OFF	ON	OFF	OFF	OFF	OFF	OFF	OFF	OFF	OFF	OFF	OFF	OFF	OFF	OFF
4	ON	OFF	OFF	OFF	OFF	ON	OFF	OFF	OFF	OFF	OFF	OFF	OFF	OFF	OFF	OFF	OFF	OFF
5	ON	OFF	OFF	OFF	OFF	OFF	ON	OFF	OFF	OFF	OFF	OFF	OFF	OFF	OFF	OFF	OFF	OFF
6	ON	OFF	OFF	OFF	OFF	OFF	OFF	ON	OFF	OFF	OFF	OFF	OFF	OFF	OFF	OFF	OFF	OFF
7	ON	OFF	OFF	OFF	OFF	OFF	OFF	OFF	ON	OFF	OFF	OFF	OFF	OFF	OFF	OFF	OFF	OFF
8	ON	OFF	OFF	OFF	OFF	OFF	OFF	OFF	OFF	ON	OFF	OFF	OFF	OFF	OFF	OFF	OFF	OFF
9	OFF	ON	OFF	OFF	OFF	OFF	OFF	OFF	OFF	OFF	ON	OFF	OFF	OFF	OFF	OFF	OFF	OFF
10	OFF	ON	OFF	OFF	OFF	OFF	OFF	OFF	OFF	OFF	OFF	ON	OFF	OFF	OFF	OFF	OFF	OFF
11	OFF	ON	OFF	OFF	OFF	OFF	OFF	OFF	OFF	OFF	OFF	OFF	ON	OFF	OFF	OFF	OFF	OFF
12	OFF	ON	OFF	OFF	OFF	OFF	OFF	OFF	OFF	OFF	OFF	OFF	OFF	ON	OFF	OFF	OFF	OFF
13	OFF	ON	OFF	OFF	OFF	OFF	OFF	OFF	OFF	OFF	OFF	OFF	OFF	OFF	ON	OFF	OFF	OFF
14	OFF	ON	OFF	OFF	OFF	OFF	OFF	OFF	OFF	OFF	OFF	OFF	OFF	OFF	OFF	ON	OFF	OFF
15	OFF	ON	OFF	OFF	OFF	OFF	OFF	OFF	OFF	OFF	OFF	OFF	OFF	OFF	OFF	OFF	ON	OFF
16	OFF	ON	OFF	OFF	OFF	OFF	OFF	OFF	OFF	OFF	OFF	OFF	OFF	OFF	OFF	OFF	OFF	ON

3. Analytical Results and Discussion

In this section, we investigated the modeled antenna, using an Ansys high frequency structure simulator (Electromagnetics Suit, HFSS, Ansys, Canonsburg, PA, USA). We analyzed the measured and simulated results for the proposed antenna with return loss (*VSWR*), as well as the gain and efficiency steering at different modes and states for both 28 GHz and 38 GHz.

3.1. Frequency/VSWR/Bandwidth/Gain/Efficiency Reconfiguration Analysis

In this subsection, the frequency reconfigurability of the proposed model 1 was observed using two switches, i.e., SD1 and SD2. In Figure 4 and Table 4, it can be seen that when SD1 was ON and the other switches were OFF, the proposed antenna resonated at 28.10 GHz (simulated) and 28.66 GHz (measured). Similarly, a frequency of 37.63 GHz (simulated) and resonant frequency of 38.3 GHz (measured) was perceived when switch SD2 was in the ON state, while the rest of the switches were kept OFF, subjected to the unidirectional mmWaves of 5G mobile communication. Further, RL < −10 dB for both 28 GHz and 38 GHz mmWave frequency bands was the finite result of the simulations measured, with a small dissimilarity. These results were due to losses connected with SMA (Sub-Miniature version A), tangent loss, improper soldering and other environmental factors arising during the fabrication process. Figure 5 shows the simulated and measured *VSWR* for every state (from Table 2). The results show that, for State 2, the impedance bands were 1.4 GHz and 1.51 GHz for 28 GHz and 38 GHz (<2.5 standards, respectively).

Table 4. Bandwidth (BW), gain, efficiency and voltage standing wave ratio (*VSWR*) analysis at different states for frequency reconfiguration.

State	BW (%)	BW (GHz)	Gain (dBi)	Efficiency (%)	*VSWR*
State 1	40.85	26.40–39.722	3.8	53.2	-
State 2	12.707	26.240–29.748	8.3	70.20	1.4
State 3	7.65	36.812–39.722	7.1	66.39	1.51

While investigating the parametric models, the right structure was chosen and then analyzed individually for the return loss of selected frequencies, rather analyzing the whole design (Figure 6).

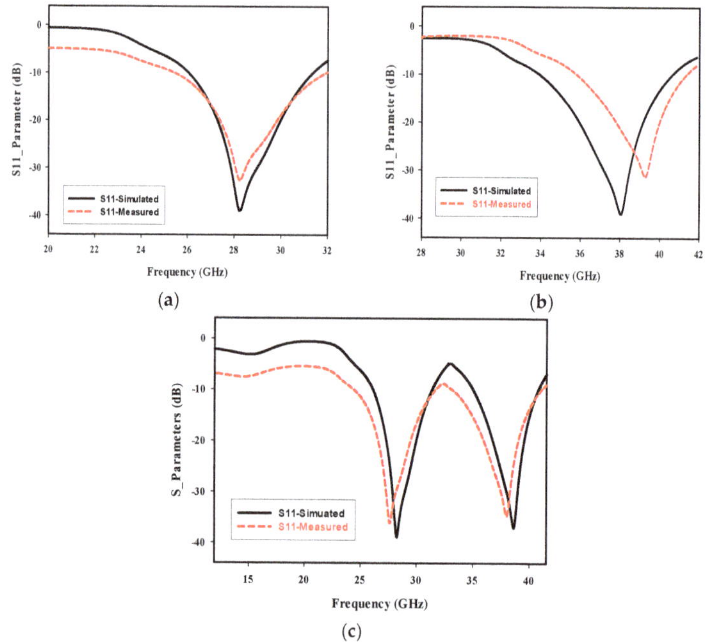

Figure 4. (**a**) State 1, (**b**) State 2 and (**c**) State 3. Return loss at different states of Table 2.

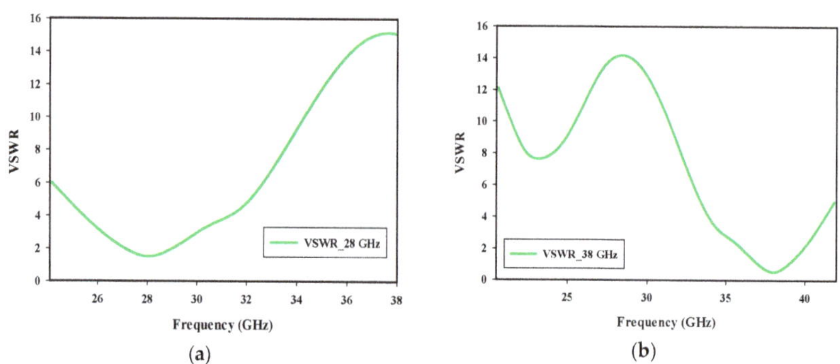

Figure 5. *VSWR* at (**a**) State 2 and (**b**) State 3 of Table 2.

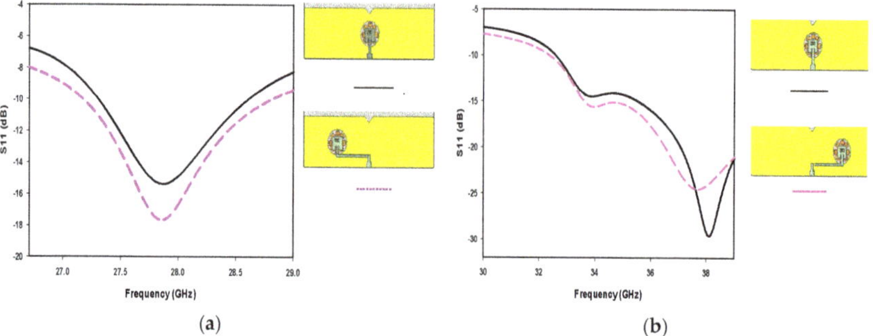

Figure 6. Parametric analysis of individual patches for (**a**) 28 GHz and (**b**) 38 GHz.

3.2. Radiation Pattern Reconfiguration Analysis

In this section, we discuss the pattern reconfiguration of different modes presented in Table 3. For pattern steering, parasitic stubs were connected with partial ground through NMOS switches (SD1–SD18). The purpose of these switches was to yield the change in surface current distribution, which would thus change the characteristics of the radiation patterning. As a result, the radiation pattern and surface current density of the antenna operating at different switching modes was derived as 45° shift along the XOY-plane (E-plane) at $O = 0°$. This direction along the XOY-plane was altered by changing the ON-OFF state in SD1–SD18. In accordance with the ON-OFF states of the parasitic stubs with symmetrical antenna structures, the single main beam's direction was symmetrically variant along the z-axis in the XOY-plane (Figure 7, from mode 1–mode 16). These modes showed narrow HPBW (Half Power Beamwidth) and a high-quality pattern.

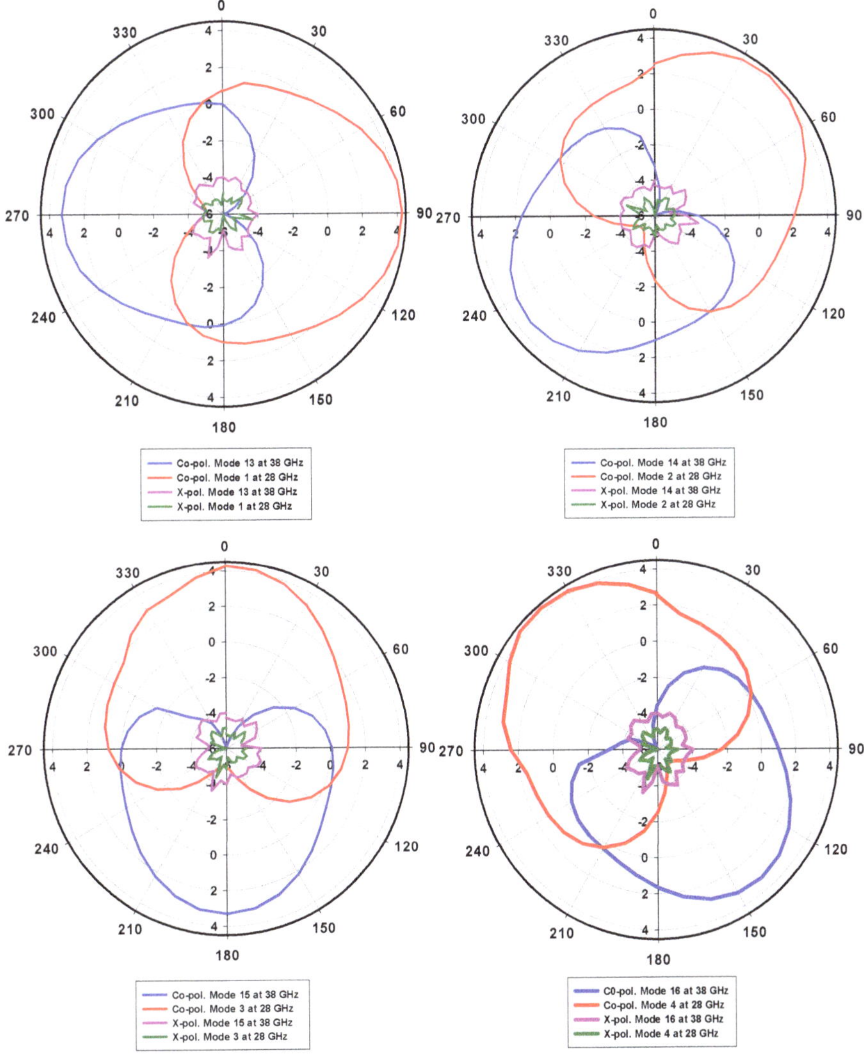

Figure 7. *Cont.*

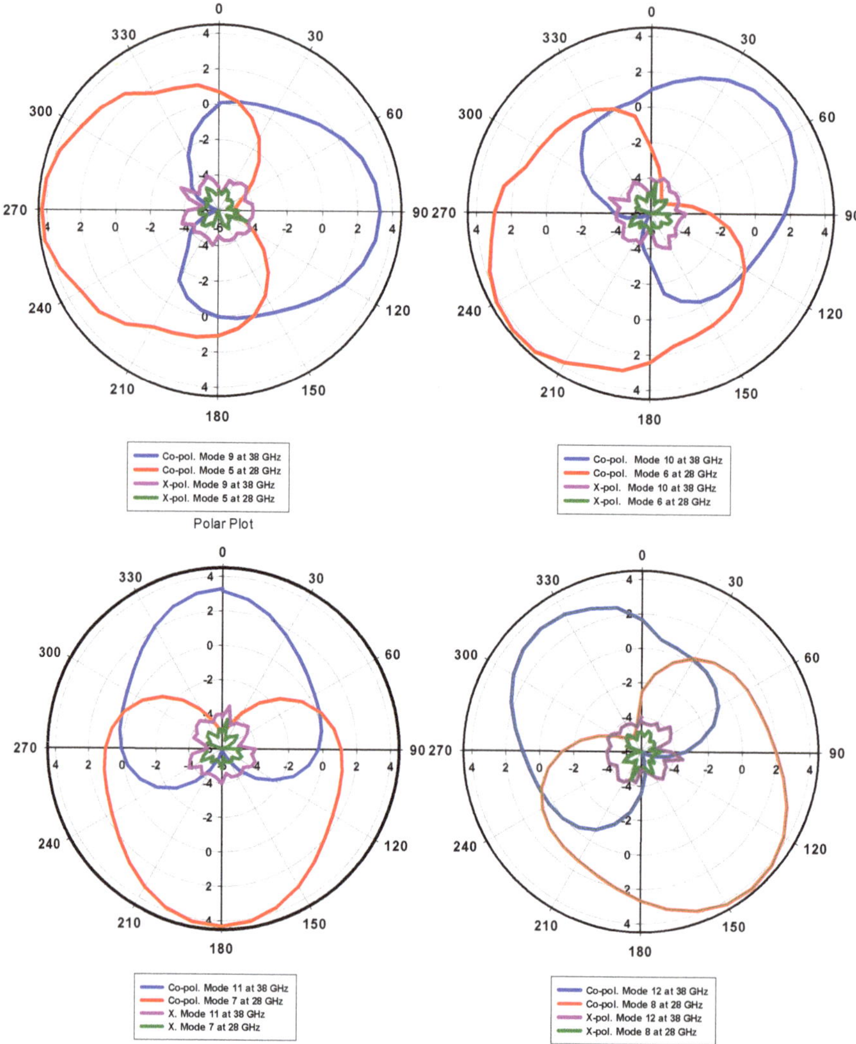

Figure 7. Co- and cross (X) polarization results showing radiation pattern of both 28 GHz and 38 GHz on XOY-E-Plane at different modes from Table 3.

Figure 7 shows the co- and cross (X) polarization radiation pattern results from Table 3 (from mode 1 to 16). It illustrates that from mode 1 to 8, SD1 is ON and SD2 is OFF (SD1 and SD2 are the main cause of frequency reconfiguration), operating with a resonant frequency of 28 GHz (it is unconcerned with 38 GHz). In this case, connecting the parasitic stubs from SD3 to SD10 assists in the radiation shift of the main lobe at the 28 GHz frequency. As you can see in Figure 7, the changing positions of the pattern (anti-clockwise) cause the surface current density to vary, due to the connecting parasitic stubs. Depending on the number of stubs, we can increase/decrease the steering angle (45° shift is used in this paper). Similarly, from mode 9 to 16, SD1 is OFF and SD2 is ON, which results in a 38 GHz resonant frequency. In this case, stubs being connected from SD11 to SD18, as shown in Figure 7, aids in the radiation pattern reconfiguration at 38 GHz. The novelty is that both the frequencies are controlled irrespective of each other.

Our investigation results were compared with the previously published literature, as shown in Table 5. The major attributes of the proposed model are as follows:

(a)   This structure was a novel design for mmWaves, never claimed before.

(b)   Frequency and pattern were independently controlled in a single antenna, showing structural novelty.

Table 5. Comparison with the previously published literature.

Reference	Reconfiguration	Switches/Types	Dimension(mm^2)	Generation	Gain/Max (dBi)	Distribution of Hybrid Reconfiguration
[35]	Frequency	3 PIN diodes	28 × 38	3G	N/A	-
[36]	Frequency	4 MEMS	46 × 25	4G	2.2	-
[37]	Pattern	5 varactors	38 × 42	4G	2.4	-
[38]	Pattern	4 PIN diodes	58 × 32	4G	3.14	-
[39]	Hybrid (frequency and pattern)	14 PIN diodes	150 × 160	4G	5.8/4.4/3.2	Combined
[40]	Hybrid (frequency and pattern)	N/A	86 × 48.3	4G	6.6/7.3	Combined
[41]	Hybrid (frequency and pattern)	5 PIN diodes	50 × 60	4G	4/3.3/4/5.2	Combined
[42]	Hybrid (frequency and pattern)	N/A	50 × 50	4G	6.4/2/2	Combined
This work	Hybrid (frequency and pattern)	18 NMOS transistors	112 × 52	5G	3.8/8.3/7.1	Independently tuned

4. Conclusions

In this paper, a combination of frequency and radiation pattern reconfiguration was presented for an antenna system designed for 5G applications and beyond. The proposed antenna was fabricated for 5G mmWave mobile communication at 28 GHz and 38 GHz. In total, 18 NMOS transistors were used as switches for the ON and OFF states. Two switches, SD1 and SD2, were used for frequency tuning, while the other switches were utilized for pattern steering at different angles, in 45° steps, with the targeted resonant frequencies (28 GHz and 38 GHz). The steering angle is totally dependent on the number of parasitic stubs mounted on the partial ground in the proposed design, and can be altered by varying the number of stubs. The proposed model is a novel structure and can be used for future 5G mobile communication applications.

Author Contributions: M.K.S. provided the idea, performed the experiments and managed the paper. M.I.K. conceived of the presented idea, verified the analytical methods contributed to the interpretation of the results and took the lead in writing the manuscript. M.A.-H. assisted in the idea development and paper writing. All authors discussed the results and contributed to the final manuscript. All authors have read and agreed to the published version of the manuscript.

Funding: This work is funded in part by the Abu-Dhabi Department of Education and Knowledge (ADEK) Award for Research Excellence 2019 under Grant AARE19-245.

Conflicts of Interest: The authors declare no conflict of interest.

References

1.   Haris, R.M.; Al-Maadeed, S. Integrating Blockchain Technology in 5G enabled IoT: A Review. In Proceedings of the 2020 IEEE International Conference on Informatics, IoT, and Enabling Technologies (ICIoT), Doha, Qatar, 2–5 February 2020; pp. 367–371. [CrossRef]

2.   Chettri, L.; Bera, R. A Comprehensive Survey on Internet of Things (IoT) Toward 5G Wireless Systems. *IEEE Internet Things J.* **2020**, *7*, 16–32. [CrossRef]

3.   Liu, Y.; Li, Y.; Ge, L.; Wang, J.; Ai, B. A Compact Hepta-Band Mode-Composite Antenna for Sub (6, 28, and 38) GHz Applications. *IEEE Trans. Antennas Propag.* **2020**, *68*, 2593–2602. [CrossRef]

4.   Shereen, M.K.; Khattak, M.I.; Witjaksono, G. A brief review of frequency, radiation pattern, polarization, and compound reconfigurable antennas for 5G applications. *J. Comput. Electron.* **2019**, *18*, 1065–1102. [CrossRef]

5.   Khattak, M.I.; Sohail, A.; Khan, U.; Ullah, Z.; Witjaksono, G. Elliptical Slot Circular Patch Antenna Array with Dual Band Behaviour for Future 5G Mobile Communication Networks. *Prog. Electromagn. Res. C* **2019**, *89*, 133–147. [CrossRef]

6.   Kumar, A.; Ansari, A.Q.; Kanaujia, B.K.; Kishor, J.; Tewari, N. Design of triple-band MIMO antenna with one band-notched characteristic. *Prog. Electromagn. Res. C* **2018**, *86*, 41–53. [CrossRef]

7. Alreshaid, A.T.; Hammi, O.; Sharawi, M.S.; Sarabandi, K. A compact millimeter-wave slot antenna array for 5G standards. In Proceedings of the 2015 IEEE 4th Asia-Pacific Conference on Antennas and Propagation (APCAP), Kuta, Indonesia, 30 June–3 July 2015; pp. 84–85. [CrossRef]

8. Ali, M.M.M.; Sebak, A.-R. Directive antennas for future 5G mobile wireless communications. In Proceedings of the General Assembly and Scientific Symposium of the International Union of Radio Science (URSI GASS), Montreal, QC, Canada, 19–26 August 2017; pp. 1–4. [CrossRef]

9. Ashraf, N.; Haraz, O.M.; Ali, M.M.M.; Ashraf, M.A.; Alshebili, S.A.S. Optimized broadband and dual-band printed slot antennas for future millimeter wave mobile communication. *Aeuinternational J. Electron. Commun.* **2016**, *70*, 257–264. [CrossRef]

10. Ali, M.M.M.; Sebak, A.-R. Design of compact millimeter wave massive MIMO dual-band (28/38 GHz) antenna array for future 5G communication systems. In Proceedings of the 2016 17th International Symposium on Antenna Technology and Applied Electromagnetics (ANTEM), Montreal, QC, Canada, 10–13 July 2016; pp. 1–2. [CrossRef]

11. Yan, K.; Yang, P.; Yang, F.; Zeng, L.; Huang, S. Eight-antenna array in the 5G smartphone for the dual-band MIMO system. In Proceedings of the 2018 IEEE International Symposium on Antennas and Propagation & USNC/URSI National Radio Science Meeting, Boston, MA, USA, 8–13 July 2018; pp. 41–42. [CrossRef]

12. Hasan, M.N.; Seo, M. Compact omnidirectional 28GHz 2×2 MIMO antenna array for 5G communications. In Proceedings of the 2018 International Symposium on Antennas and Propagation (ISAP), Busan, Korea, 23–26 October 2018; pp. 1–2.

13. Hong, W.; Baek, K.-H.; Ko, S. Millimeter-wave 5G antennas for smartphones: Overview and experimental demonstration. *IEEE Trans. Antennas Propag.* **2017**, *65*, 6250–6261. [CrossRef]

14. Christodoulou, C.G.; Tawk, Y.; Lane, S.A.; Erwin, S.R. Reconfigurable Antennas for Wireless and Space Applications. *Proc. IEEE* **2012**, *100*, 2250–2261. [CrossRef]

15. Freeman, J.L.; Lamberty, B.J.; Andrews, G.S. Optoelectronically reconfigurable monopole antenna. *Electron. Lett.* **1992**, *28*, 1502–1503. [CrossRef]

16. Panagamuwa, C.J.; Chauraya, A.; Vardaxoglou, J.C. Frequency and beam reconfigurable antenna using photoconducting switches. *IEEE Trans. Antennas Propag.* **2006**, *54*, 449–454. [CrossRef]

17. Roscoe, D.J.; Shafai, L.; Ittipiboon, A.; Cuhaci, M.; Douville, R. Tunable dipole antennas. In Proceedings of the IEEE/URSI International Symposium on Antennas and Propagation, Ann Arbor, MI, USA, 28 June–2 July 1993; pp. 672–675. [CrossRef]

18. Weedon, W.; Payne, W.; Rebeiz, G.; Herd, J.; Champion, M. MEMS-switched reconfigurable multi-band antenna: Design and modeling. In Proceedings of the IEEE Antennas and Propagation Society International Symposium. 1999 Digest. Held in conjunction with: USNC/URSI National Radio Science Meeting (Cat. No.99CH37010), Orlando, FL, USA, 11–16 July 1999; Volume 1, pp. 203–231.

19. Brown, E.R. On the gain of a reconfigurable-aperture antenna. *IEEE Trans. Antennas Propag.* **2001**, *49*, 1357–1362. [CrossRef]

20. Ali, M.A.; Wahid, P. A reconfigurable Yagi array for wireless applications. In Proceedings of the IEEE/URSI International Symposium on Antennas and Propagation, San Antonio, TX, USA, 16–21 June 2002; pp. 466–468. [CrossRef]

21. Vinoy, K.; Varadan, V. Design of reconfigurable fractal antennas and RF-MEMS for space-based systems. *Smart Mater. Struct.* **2001**, *10*, 1211–1223. [CrossRef]

22. Anagnostou, D.E.; Zheng, G.; Chryssomallis, M.T.; Lyke, J.C.; Ponchak, G.E.; Papapolymerou, J.; Christodoulou, C.G. Design, fabrication, and measurements of an RFMEMS-based self-similar reconfigurable antenna. *IEEE Trans. Antennas Propag.* **2006**, *54*, 422–432. [CrossRef]

23. Clarricoats, P.J.B.; Zhou, H. The design and performance of a reconfigurable mesh reflector antenna. *IEE Digit. Libr.* **1991**, *138*, 485–492.

24. Clarricoats, P.J.B.; Zhou, H.; Monk, A. Electronically controlled reconfigurable reflector antenna. In Proceedings of the Antennas and Propagation Society Symposium 1991 Digest, London, ON, Canada, 24–28 June 1991; pp. 179–181. [CrossRef]

25. Washington, G.; Yoon, H.S.; Angelino, M.; Theunissen, W.H. Design, modeling, and optimization of mechanically reconfigurable aperture antennas. *IEEE Trans. Antennas Propag.* **2002**, *50*, 628–637. [CrossRef]

26. Nikolaou, S.; Bairavasubramanian, R.; Lugo, C., Jr.; Carrasquillo, I.; Thompson, D.C.; Ponchak, G.E.; Papapolymerou, J.; Tentzeris, M.M. Pattern and frequency reconfigurable annular slot antenna using PIN diodes. *IEEE Trans. Antennas Propag.* **2006**, *54*, 439–448. [CrossRef]

27. Huff, G.H.; Feng, J.; Zhang, S.; Bernhard, J.T. A novel radiation pattern and frequency reconfigurable single turn square spiral microstrip antenna. *IEEE Microw. Wirel. Compon. Lett.* **2003**, *13*, 57–59. [CrossRef]

28. Yang, X.-S.; Wang, B.Z.; Wu, W.; Xiao, S. Yagi patch antenna with dual-band and pattern reconfigurable characteristics. *IEEE Antennas Wirel. Propag. Lett.* **2007**, *6*, 168–171. [CrossRef]

29. Ali, M.; Sayem, A.T.M.; Kunda, V.K. A reconfigurable stacked microstrip patch antenna for satellite and terrestrial links. *IEEE Trans. Veh. Technol.* **2007**, *56*, 426–435. [CrossRef]

30. Qamar, F.; Siddiqui, M.H.S.; Dimyati, K.; Noordin, K.A.B.; Majed, M.B. Channel characterization of 28 and 38 GHz MM-wave frequency band spectrum for the future 5G network. In Proceedings of the 2017 IEEE 15th Student Conference on Research and Development (SCOReD), Putrajaya, Malaysia, 13–14 December 2017; pp. 291–296. [CrossRef]

31. Stutzman, W.L.; Thiele, G.A. *Antenna Theory and Design*; John Wiley & Sons: Hoboken, NJ, USA, 2012.

32. Yasir, I.A.A.; Hasanain, A.H.A.; Baha, A.S.; Parchin, N.O.; Ahmed, M.A.; Abdulkareem, S.A.; Raed, A.A. *New Radiation Pattern-Reconfigurable 60-GHz Antenna for 5G Communications*; IntechOpen: London, UK, 2019. [CrossRef]

33. Robert, E.C. *Foundation for Microwave Engineering*, 2nd ed.; John Wiley & Sons: Hoboken, NJ, USA, 1966.

34. George, H.; Joseph, C. *Practical Antenna Handbook 5/e*; McGraw-Hill: New York, NY, USA, 2011.

35. Chitra, R.J.; Nagarajan, V. Frequency reconfigurable antenna using PIN diodes. In Proceedings of the 2014 Twentieth National Conference on Communications (NCC), Kanpur, India, 28 February–2 March 2014; pp. 1–4. [CrossRef]

36. Yamagajo, T.; Koga, Y. Frequency reconfigurable antenna with MEMS switches for mobile terminals. In Proceedings of the 2011 IEEE-APS Topical Conference on Antennas and Propagation in Wireless Communications, Torino, Italy, 12–16 September 2011; pp. 1213–1216. [CrossRef]

37. Tian, H.; Jiang, L.J.; Itoh, T. A Compact Single-Element Pattern Reconfigurable Antenna with Wide-Angle Scanning Tuned by a Single Varactor. *Prog. Electromagn. Res. C* **2019**, *92*, 137–150. [CrossRef]

38. Kang, W.; Lee, S.; Kim, K. A pattern-reconfigurable antenna using PIN diodes. *Microw. Opt. Technol. Lett.* **2011**, *53*, 1883–1887. [CrossRef]

39. Li, W.; Ren, Z.; Shi, X.; Hei, Y. A frequency and pattern reconfigurable microstrip antenna using PIN diodes. In Proceedings of the IEEE Antennas and Propagation Society International Symposium (APSURSI), Memphis, TN, USA, 6–11 July 2014; pp. 1449–1450. [CrossRef]

40. Ghaffar, A.; Li, X.J.; Hussain, N.; Awan, W.A. Flexible Frequency and Radiation Pattern Reconfigurable Antenna for Multi-Band Applications. In Proceedings of the 2020 4th Australian Microwave Symposium (AMS), Sydney, Australia, 13–14 February 2020; pp. 1–2. [CrossRef]

41. Li, P.K.; Shao, Z.H.; Wang, Q.; Cheng, Y.J. Frequency- and pattern-reconfigurable antenna for multistandard wireless applications. *IEEE Antennas Wirel. Propag. Lett.* **2015**, *14*, 333–336. [CrossRef]

42. Majid, H.A.; Rahim, M.K.A.; Hamid, M.R.; Ismail, M.F. Frequency and pattern reconfigurable slot antenna. *IEEE Trans. Antennas Propag.* **2014**, *62*, 5339–5343. [CrossRef]

 *electronics*

Article

Ultra-Compact Reconfigurable Band Reject UWB MIMO Antenna with Four Radiators

Muhammad Saeed Khan [1],[†] , Adnan Iftikhar [1], Raed M. Shubair [2],[3],
Antonio-Daniele Capobianco [4], Sajid Mehmood Asif [5] and Benjamin D. Braaten [6]
and Dimitris E. Anagnostou [7],*

1   Department Electrical and Computer Engineering, COMSATS University Islamabad,
    Islamabad 45550, Pakistan; mskj786@hotmail.com (M.S.K.); adnaniftikhar@comsats.edu.pk (A.I.)
2   Research Laboratory of Electronics, Massachusetts Institute of Technology (MIT),
    Cambridge, MA 02139, USA; rshubair@mit.edu
3   Department of Electrical and Computer Engineering, New York University, Abu Dhabi 129188, UAE;
    raed.shubair@nyu.edu
4   Department of Information Engineering, University of Padova, Via Gradenigo 6/b, 35131 Padova, Italy;
    adc@dei.unipd.it
5   Electronic and Electrical Engineering, The University of Sheffield, Sheffield S1 4ET, UK; sajid.asif@gmail.com
6   Department of Electrical and Computer Engineering, North Dakota State University, Fargo, ND 58102, USA;
    benjamin.braaten@ndsu.edu
7   Department of Electrical Engineering, Heriot-Watt University, Edinburgh EH14 4AS, UK
*   Correspondence: d.anagnostou@hw.ac.uk
†   These authors contributed equally to this work.

Received: 19 February 2020; Accepted: 23 March 2020; Published: 30 March 2020

Abstract: A compact reconfigurable UWB MIMO antenna with four radiators that accomplish on-demand band rejection from 4.9 to 6.3 GHz is presented. An LC stub is connected to the ground plane by activating the PIN diode for each radiator. Two radiators are placed perpendicular to each other to exploit the polarization diversity on a compact 25×50 mm^2 FR4 laminate. Two additional radiators are then fixed obliquely on the same laminate (without increasing size) in angular configuration at $\pm 45°$ perpendicular to the first two planar radiators still exploiting polarization diversity. The design is validated by prototyping and comparing the results with the simulated ones. On demand band rejection through the use of PIN diodes, wide impedance matching (2–12 GHz), high isolation amongst the radiators, compactness achieved by angular placement of the radiators, low gain variation over the entire bandwidth, band rejection control achieved by adjusting the gap between stub and ground plane, and low TARC values makes the proposed design very suitable for commercial handheld devices (i.e., Huawei E5785 and Netgear 815S housings). The proposed configuration of the UWB MIMO radiators has been investigated first time as per authors' knowledge.

Keywords: band rejected; envelope correlation co-efficient; four element MIMO; polarization diversity; ultra-wideband multiple input multiple output

---

1. Introduction

For the present and future wireless technologies, Wireless Personal Area Networks (WPANs), Ultra-wideband (UWB), and Multiple Input Multiple Output (MIMO) antennas are fundamental and indispensable transducers. These transducers enhance MIMO system performance using antenna diversity techniques, such as spatial diversity, pattern diversity, or polarization diversity. Nonetheless, unwanted interference from Wireless Local Area Networks (WLAN), i.e., IEEE 802.11 (5.8 GHz), may cause malfunctioning or weak precision in any UWB MIMO transceiver for the UWB application such as sensor data collection, precision locating, or tracking applications [1–5]. Therefore, MIMO antenna designers and researchers have proposed band-reject UWB MIMO antennas to mitigate WLAN interference [6–10].

The use of integrated elements in the feed line can result in multiple band notch characteristics in a UWB antenna, as reported in [6]. The etching of slots in the radiator can also notch dual band in a UWB MIMO antenna [7]. Similarly, introduction of slits in an antenna size of 36×36 mm^2 can be used to achieve band notch functionality [8]. In addition, inclusion of stubs in the ground plane can notch a specific band over a UWB spectrum, but this technique results in an overall antenna size of 40×68 mm^2, as reported in [9]. Further, the incorporation of $\lambda/2$ open stubs on the antenna having ground plane size of 55×86.5 mm^2 successfully achieved a notched band; however, the use of large reflectors restrict further size reduction of the antenna reported in [10]. Likewise, the work presented in [11–13] highlights the use of stubs for WLAN band rejection along with the exploitation of polarization diversity for isolation enhancement amongst the elements. Loading of radiator or ground plane with metamaterials can also notch a WLAN band. In this regard, the authors of [14] implemented an Electromagnetic bandgap (EBG) structure for band rejection and a band stop filter for improving coupling. Huang et al. [15] proposed Complementary Split Rings Resonator (CSRR) slots and open stubs near the feedline to achieve band notch functionality. Polarization diversity in the design was attained by vertical placement of the designed antennas [15]. Moreover, defected ground plane in circular monopoles was used to notch 4.6 GHz band, and periodic EBG structures with large number of vias were placed between 4×4 MIMO antennas to achieve isolation of more than 17 dB [16]. Similarly, a planar placement of 4×4 UWB MIMO antenna having isolation of more than 17 dB was presented in [17]. A four element UWB MIMO antenna with a complex band stop design for the WLAN band resulted in an overall footprint of 39.8×50 mm^2 [18], whereas a similar placement of four element UWB antenna having modified ground plane with slots had an overall size of 45×45 mm^2 [18]. It can be seen from the aforementioned discussion that researchers have successfully addressed different techniques to notch a single band [8–10] or multiple bands [6,7] in UWB MIMO antennas. However, these techniques permanently reject a specific band and thus provide inflexibility in the antenna design to un-notch that specific band. Moreover, most of the antenna designs presented in the literature deliver band rejection in a two elements UWB antenna system.

Conversely, on demand rejection using active switches such as PIN diodes, RF microelectromechanical system (MEMS) switches, and optical switches have also been proposed in the literature [19–24]. However, the demonstrations of reconfiguration mechanism/on-demand band rejection has only been performed on a single or at most of two elements antenna system [22]. Therefore, overall system performance of a four-element UWB MIMO antenna having on demand band rejection capabilities cannot be accurately estimated by exploring two element UWB MIMO designs. Besides, a major portion of the literature is focused on the right angled placement of the MIMO antenna elements, which occupies a major portion of the device housing. The right angle placement of UWB MIMO antenna elements occupies extra space and can therefore be unfit for modern 4G LTE WiFi hotspots such as Huawei E5785 and Netgear 815S. Therefore, the objective of this work is twofold: (a) introduce an efficient method of utilizing UWB spectrum by proposing on-demand band rejection through the use of PIN diode;s and (b) present a novel design of a four-element UWB MIMO antenna system which makes use of two slanted radiators placed at an angle of $\pm 45°$; this leads to the capability of on demand band rejection and space saving of device housing.

In this paper, an ultra-compact frequency reconfigurable UWB MIMO antenna is presented. All four radiators are capable of rejecting the WLAN signals on demand by activating the PIN diodes. PIN diodes connect the $\lambda_g/4$ stub to the ground plane to draw most of the current on the stub. Two radiators are placed on the planar substrate and two radiators are placed angularly at $\pm 45°$ perpendicular to the planar laminate, as shown in Figure 1. The overall size in the planar configuration is only 25×50 mm^2 and specific angular placement of the radiators provide liberty to the designer to use housing height to adjust the radiators. In addition, adding reconfigurability in the proposed design makes it valuable for most of the UWB MIMO applications. The proposed method to reduce the overall size of the UWB MIMO antenna, to the best of authors' knowledge, has never been investigated before. The use of disconnected ground planes provides fabrication simplicity and mitigate coupling issues in such a closely packed configuration. On the other hand, the incorporation of decoupling structures may replace separate ground planes requirement but at the cost of increased hardware complexity and fabrication challenges [23,24].

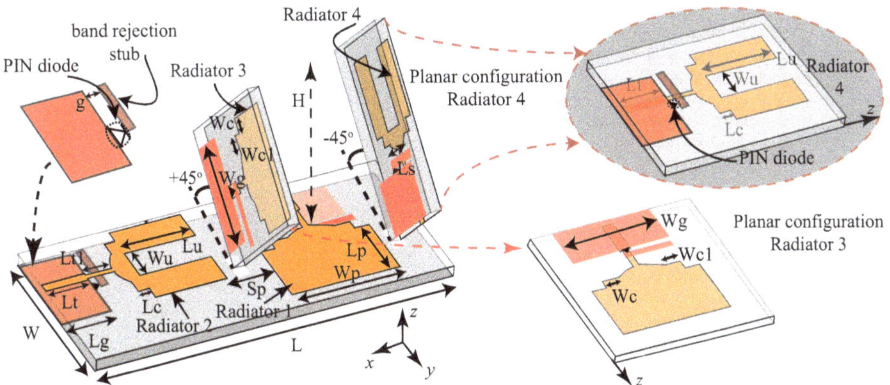

Figure 1. Perspective view of the proposed UWB MIMO antenna (dimensions in mm): L = 50, W = 25, g = 0.5, L_p = 10, W_p = 15, W_{p1} = 5, W_u = 5, L_u = 10, L_c = 1.5, W_c = 2, W_{c1} = 2.25, L_t = 6, L_{t1} = 3.82, W_g = 13.5, L_g = 7, L_s = 7.25, H = 22, and S_p = 6.15. Radiators 1 and 2 are identical in shape and size to radiators 3 and 4, respectively.

2. Design Procedure

The design procedure is started by optimizing the single element of the proposed UWB MIMO antenna for the complete 2–12 GHz frequency range. Multiple techniques have been integrated to achieve the final optimum design which includes impedance transformer with a transmission line for lower band matching (3–5 GHz), tapered section at the input of radiator for matching around 7 GHz and corner truncation of the rectangular patch for the higher band (10–12 GHz). On the bottom side of the radiator, a stub is connected with the partial ground plane to reject the WLAN band. Later, to achieve the reconfigurability of the WLAN band, the stub is connected or disconnected through PIN diodes. When the PIN diode is turned on, the length of the stub acts an inductor, while the gap between the stub and ground plane act as a capacitor, resulting in a LC band-stop filter. The length of the stub is calculated using $L_s = \dfrac{\lambda_g}{4} = \dfrac{c}{4f_o\sqrt{\epsilon_r}}$, where $f_o \approx 5$ GHz, and ϵ_r is the relative permittivity of the substrate. Next, a second element (identical to first element) is placed on the same laminate at a distance of 6.15 mm and rotated by 90°. Due to small spacing, the impedance matching is affected: to improve the input impedance, the second element is then modified by inserting a U-shaped slot. The length and width of U-slot are optimized using $L_u = \dfrac{\lambda_g}{4} = \dfrac{c}{4f_o\sqrt{\epsilon_r}}$ and $W_u = \dfrac{\lambda_g}{8} = \dfrac{c}{8f_o\sqrt{\epsilon_r}}$, where $f_o \approx 3.65$ GHz, and ϵ_r is the relative permittivity of the substrate. The comparison of reflection co-efficient with and without U-slot is shown in Figure 2.

As can be seen in Figure 2, the reflection co-efficient of second element is improved when inserting the U-slot. It can be observed that Radiators 1 and 2 are almost identical, the only modification in Radiator 2 being the U-shaped slot. All other parameters of radiators, ground plane, and stubs are exactly same as those for Radiator 1. Now, to add more elements in the design, without increasing further the dimensions of the laminate while keeping low the mutual coupling, two additional radiators are placed perpendicularly to the substrate and slanted in angular configuration at ±45°. The specific angle was chosen after several numerical iterations to have low mutual coupling and polarization diversity between Radiators 3 and 4. In the proposed design, the height of the design is 22 mm.

This allows utilizing the height of the housing, which increases from 20 to 80 mm according to the application and design (i.e., Huawei E5785 and Netgear 815S housings). Radiator 3 is identical to Radiator 1 and Radiator 4 is identical to Radiator 2. The final layout of the proposed UWB MIMO antenna system is shown in Figure 1. All EM simulations and design optimization were carried out in ANSYS HFSS.

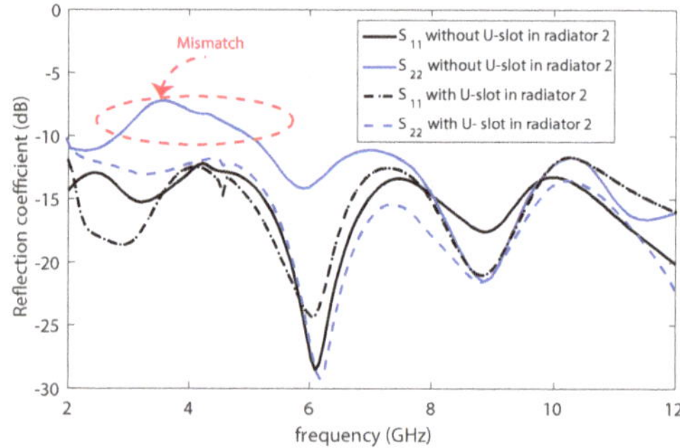

Figure 2. Comparison of the simulated reflection coefficient of Radiators 1 and 2 with and without U-slot. A mismatch is observed at lower frequencies, when U-slot is not inserted in one of the radiators.

3. Results and Discussion

3.1. S Parameters

To validate the performance of the proposed antenna, a prototype of the design reported in Figure 1 was fabricated on a 1.6 mm thick FR4 substrate with a relative permittivity of 4.5 and a loss tangent of 0.02. The fabricated prototype is shown in Figure 3. Surface mount voltage controlled PIN diodes were used to connect the stub to the ground plane and control voltage ($+V = 0.7$ V) was applied on the conducting surface with RF chokes. ANSYS HFSS was used to model the PIN diodes using lumped elements in the simulation and, during measurement, DC biased signal was used to control the ON and OFF state of the PIN diodes.

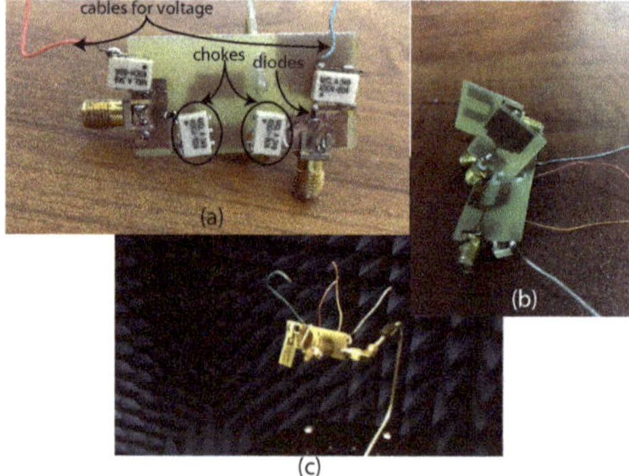

Figure 3. Fabricated prototype with dimensions shown in Figure 1: (**a**) back side view with PIN diodes and RF chokes; (**b**) perspective view; and (**c**) antenna mounted in chamber for pattern measurements.

The reflection coefficient and mutual coupling for 'ON' (i.e., all four diodes biased, and the stubs physically connected with the ground) and 'OFF' (i.e., all four diodes unbiased, and the stubs physically disconnected from the ground) states were measured using an Agilent N5242A PNA-X network analyzer. These results are shown in Figures 4–7. Figure 4a,b shows the comparison of simulated and measured reflection coefficient of Elements 1 and 2 and Elements 3 and 4, respectively, when diodes are OFF. The measured reflection coefficient is better than -10 dB on all ports over the entire band and a good agreement can be seen. The mutual coupling amongst elements is less than -17 dB over the whole band. The simulated and measured mutual couplings are plotted in Figure 5a,b, respectively. Similarly, when diodes are turned ON, measured results regarding reflection coefficient and mutual coupling are plotted in Figures 6 and 7. A strong band rejection was observed from 4.9 to 6.3 GHz.

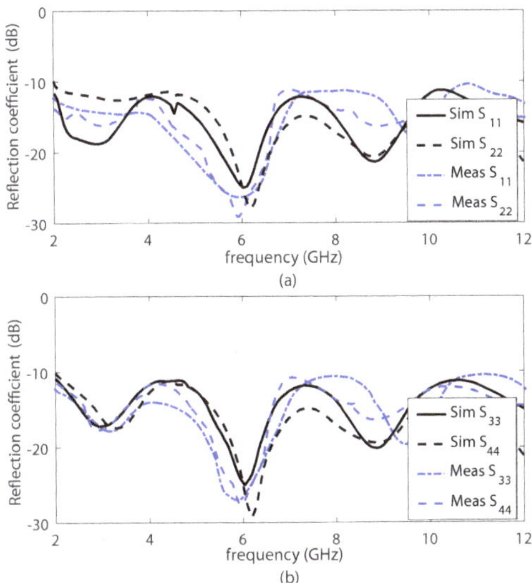

Figure 4. Simulated and measured reflection coefficient below -10 dB, when diodes are turned OFF: (**a**) comparison of simulated and measured reflection coefficients at Ports 1 and 2; and (**b**) comparison of simulated and measured reflection coefficients at Ports 3 and 4.

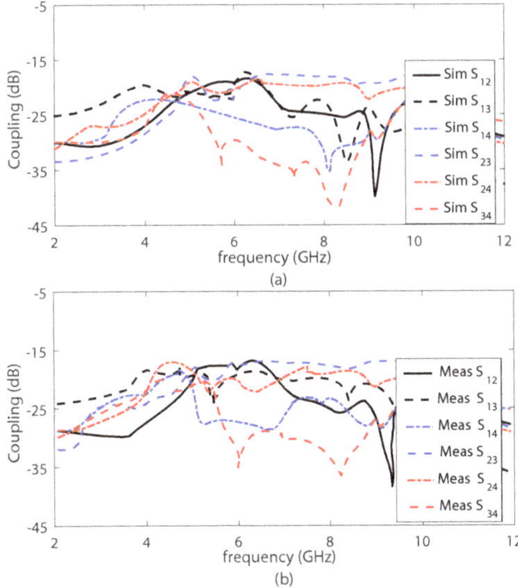

Figure 5. Simulated and measured mutual couplings less than -17 dB amongst all ports, when diodes are turned OFF: (**a**) simulated; and (**b**) measured only.

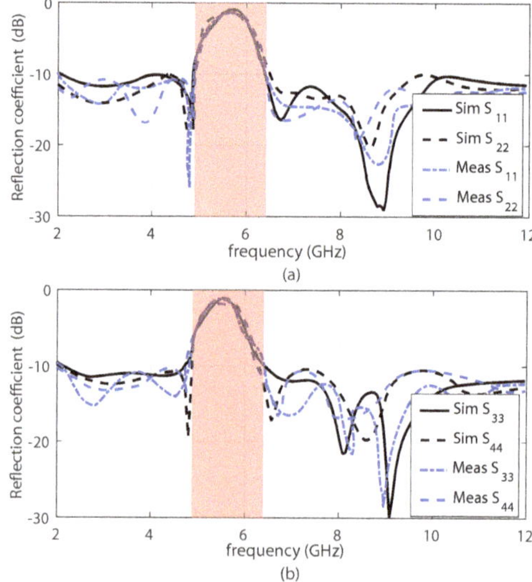

(a)

(b)

Figure 6. Simulated and measured reflection coefficient below −10 dB except the rejected band, where values reach −1 dB almost at all input ports, when diodes are turned ON (**a**) comparison of simulated and measured reflection coefficients at Ports 1 and 2; and (**b**) comparison of simulated and measured reflection coefficients at Ports 3 and 4.

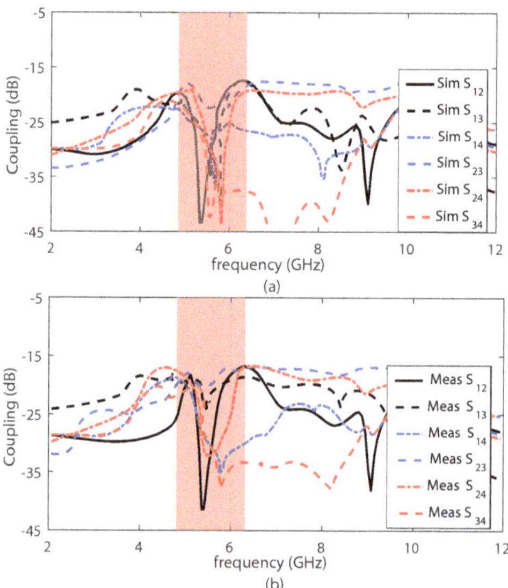

(a)

(b)

Figure 7. Simulated and measured mutual couplings less than −17 dB amongst all ports, when diodes are turned ON: (**a**) simulated; and (**b**) measured only.

3.2. Radiation Pattern and Peak Gain

The radiation patterns at three different frequencies (3.5, 5.8, and 9.8 GHz) for ON and OFF state were measured in the anechoic chamber in the principle x-z and y-z planes, and compared with the simulated patterns. During the measurements, all ports were terminated with a 50-Ω matched load except the Port 1. Only slight variations in the patterns were observed outside the rejected band during the ON and OFF state of diodes. Patterns at lower frequencies for both ON and OFF state are fairly dumbbell-shaped in the y-z plane and omnidirectional in the x-z plane, as shown in Figure 8a,b; a good agreement with simulated patterns was observed. At 5.8 GHz, when diodes are turned ON, the antenna radiates with very low intensity, which is evident in Figure 8c,d, while the antenna radiates with high intensity when diodes are turned OFF. At higher frequencies (9.8 GHz), as shown in Figure 8e,f, patterns are still omnidirectional in the x-z plane and slightly deviate from dumbbell-shaped. The discrepancy from simulated results is more evident, due to fabrication tolerances, soldering effects, cable, and connector losses and not ideal PIN diodes and RF chokes.

To check the gain variation, the peak gain over the entire bandwidth was measured for Port 1, as plotted in Figure 9 along with the simulated one for both ON and OFF state. It is evident in Figure 9 that gain lowers to -4 dBi in the rejected band, showing that radiation intensity is very low in this band when diode is turned ON, while the gain rises up to 3.8 dB when diodes are turned OFF. The peak gain variation is less than 3.5 dBi in the entire band while diode is turned OFF.

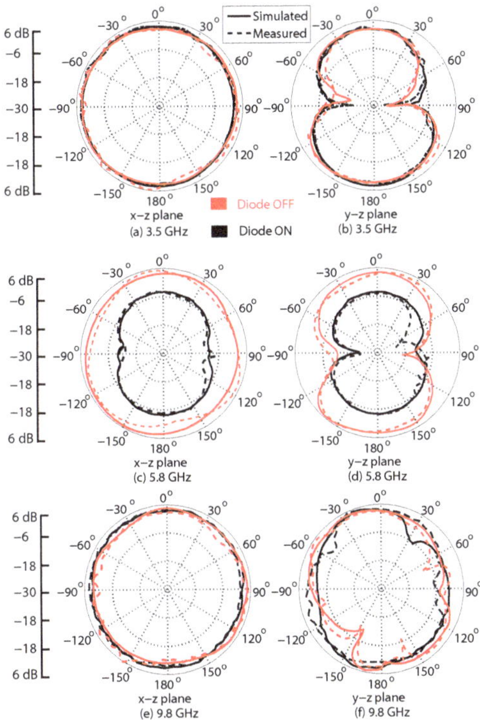

Figure 8. Simulated and measured radiation patterns in the principle planes, when diodes are turned OFF (red color) and turned ON (black color): (**a**) x-z plane at 3.5 GHz; (**b**) y-z plane at 3.5 GHz; (**c**) x-z plane at 5.8 GHz; (**d**) y-z plane at 5.8 GHz; (**e**) x-z plane at 9.8 GHz; and (**f**) y-z plane at 9.8 GHz. The patterns are nearly omni-directional in the x-z plane, suitable for UWB MIMO systems.

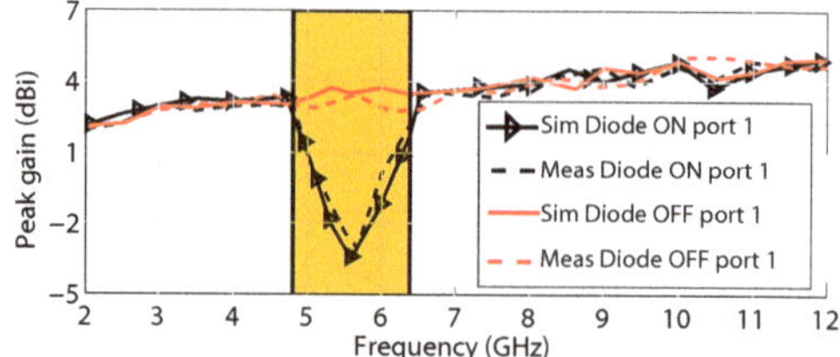

Figure 9. Simulated and measured peak gain over the entire band, when diodes are turned ON and OFF; the gain varies from 2.0 to 5.5 dBi except the rejected band during ON state, where gain drops down to −4 dBi.

3.3. Diversity Analysis

In MIMO systems, the multipath effects can be mitigated, if two or more elements have different patterns in a particular plane. In the proposed design, Radiators 1 and 2 are placed perpendicular, so they have different patterns in the respective planes. The x-z patterns of Radiator 1 are almost identical to the y-z patterns of Radiator 2. Similarly, Radiators 3 and 4 have pattern diversity, as shown in Figure 10 by plotting 3D radiation patterns at 4 GHz. It can be seen that null is located at 45° when Port 3 is excited and −45° when Port 4 is excited. This behavior can also be visualized by plotting the 2-D radiation patterns of Radiators 3 and 4 in the y-z plane, as shown in Figure 11. It is evident that the simulated and measured patterns of Radiators 3 and 4 are un-correlated and almost orthogonal. In fact, Radiator 3 has a maximum radiation at −45° and 135°, while Radiator 4 has null in those directions. Similarly, Radiator 4 has a maximum radiation at 45° and −135°, while Radiator 3 has null in those directions.

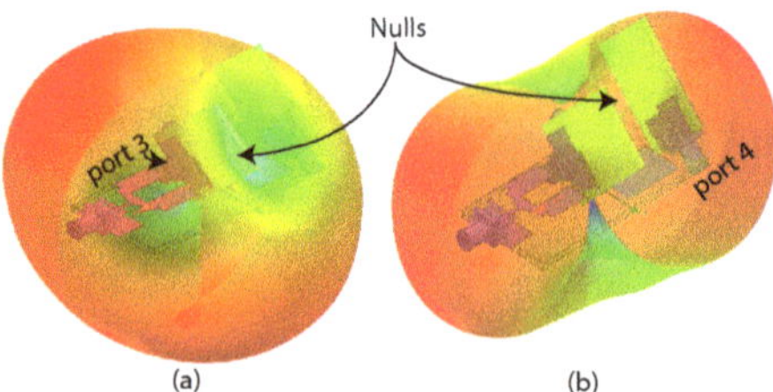

Figure 10. Simulated 3D radiation patterns at 4 GHz showing the pattern diversity for Ports 3 and 4: (**a**) only Port 3 is excited; and (**b**) only Port 4 is excited.

The un-correlated (orthogonal) radiation patterns, giving rise to diversity which is exploited in MIMO technology, are beneficial in a rich multipath environment. To check the performance of the proposed antenna in a complex scenario, envelope correlation coefficient (ECC) was calculated from the far-field radiation patterns. For different environments, such as indoor and outdoor, the parameters calculated in [25] are used in Equation (1) and numerically calculated values of ECC are plotted in Figure 12 for both ON and OFF state. The computed values from far-field radiation patterns are less than 0.43 for all cases. For uniform scattering environments, the values are less than 0.15.

$$\rho_e = \frac{\left| \int_0^{2\pi} \int_0^{\pi} \left(XPR.E_{\theta 1}.E_{\theta 2}^*.P_\theta + E_{\varphi 1}.E_{\varphi 2}^*.P_\varphi \right) d\Omega \right|^2}{\int_0^{2\pi} \int_0^{\pi} \left(XPR.E_{\theta 1}.E_{\theta 1}^*.P_\theta + E_{\varphi 1}.E_{\varphi 1}^*.P_\varphi \right) d\Omega \times \int_0^{2\pi} \int_0^{\pi} \left(XPR.E_{\theta 2}.E_{\theta 2}^*.P_\theta + E_{\varphi 2}.E_{\varphi 2}^*.P_\varphi \right) d\Omega} \tag{1}$$

In MIMO systems, when multiple antennas are actively involved in transmission or reception, they change the efficiency and overall operating bandwidth. Therefore, total active reflection coefficient (TARC) should be lower than 0 dB, which incorporates overall combined effect of all antenna in a

MIMO system and is determined by Equation (2). TARC is plotted in Figure 13 for both ON and OFF states and is less than −9 dB over the entire band.

$$TARC = \sqrt{\frac{|S_{11}+S_{12}+S_{13}+S_{14}|^2+|S_{21}+S_{22}+S_{23}+S_{24}|^2+|S_{31}+S_{32}+S_{33}+S_{34}|^2+|S_{41}+S_{42}+S_{43}+S_{44}|^2}{4}} \qquad (2)$$

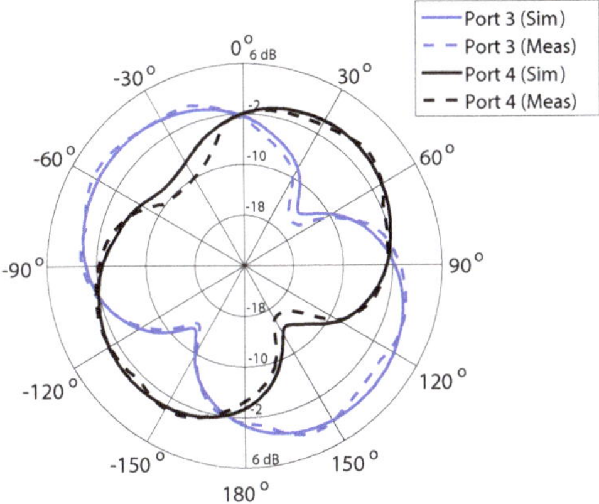

Figure 11. Simulated and measured radiation patterns of Radiators 3 and 4 in the y-z plane at 4 GHz plane showing strong un-correlation at some angles and almost orthogonal, useful for diversity applications.

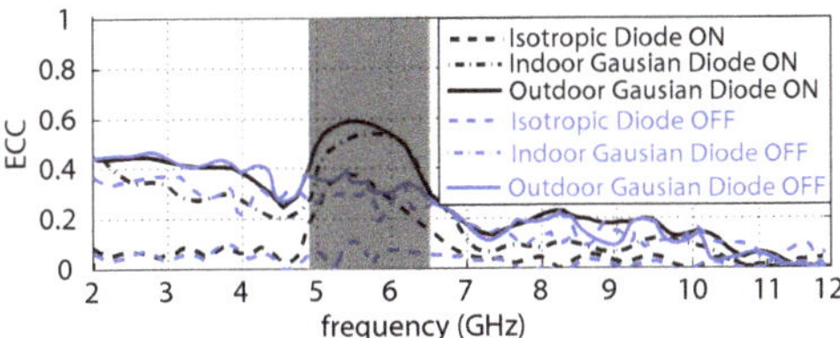

Figure 12. Computed envelope correlation coefficients between Radiators 3 and 4 from far-field radiation pattern for isotropic (uniform), indoor and outdoor environments. The XPR values used for indoor and outdoor environment are 5 and 1 dB, respectively. The ECC values are less than 0.43 in all cases.

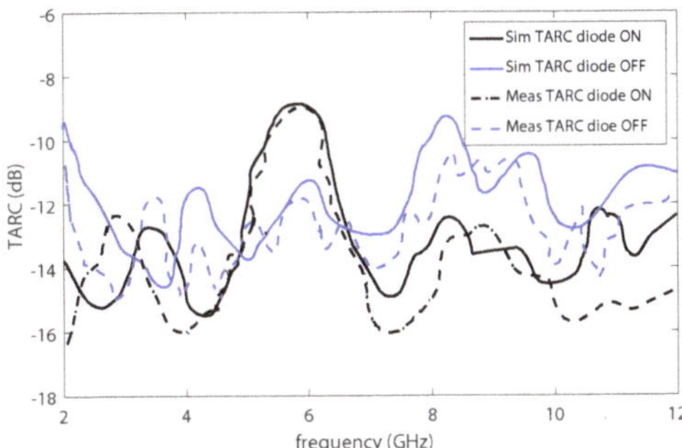

Figure 13. Simulated and measured TARC for both ON and OFF state calculated from S-parameters. TARC is lower than −9 dB over the entire band.

3.4. Parametric Analysis on Bandwidth Control

One of the main advantages of the proposed design is that the rejected frequency region can be controlled by changing the gap (g) between ground plane and stub. To verify this, the gap was varied from 0.25 to 1.5 mm and results are plotted in Figure 14. The variation in the gap changes the capacitance between ground plane and stub, resulting in the change of rejected bandwidth. When the gap is 0.25 mm, the rejected band is from 5.25 to 6.25 GHz, and it keeps on increasing while gap is increased. The maximum bandwidth of the rejected band can be obtained around 2.6 GHz (3.7–6.3 GHz), when the gap is 1.5 mm.

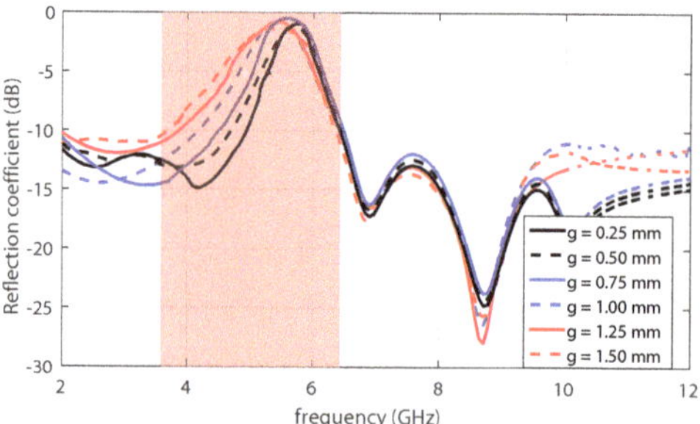

Figure 14. Effects of the gap variation between ground plane and stub on reflection coefficients. While the values are changed from 0.25 to 1.50 mm, the rejected bandwidth can be controlled from 1 to 2.6 GHz.

To further investigate the effect of the stub on the antenna, the surface currents are plotted in Figure 15 for both ON and OFF states at 5.8 GHz. It is evident in Figure 15a that stub is not affected when the diode is turned OFF as very little current is coupled on it, while turning the diode ON, high currents flow on the stub (Figure 15b), thus rejecting the band.

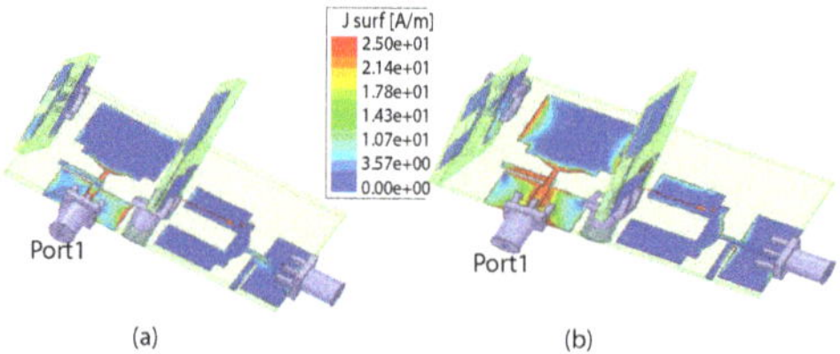

Figure 15. Surface current distribution for OFF and ON state: (**a**) exciting Port 1 at 5.8 GHz, while diode is OFF; and (**b**) exciting Port 1 only at 5.8 GHz, while diode is turned ON. The plot illustrates that stub only has current on it in the rejected band, when diode is turned ON.

3.5. Comparison With Previous Work Done

In Table 1, the proposed antenna is compared with the most representative UWB MIMO antennas existing in the literature. This list in not complete but offers a fair insight on the current work. Some of the existing antennas are compact but do not provide the band notch design; some compact designs provide the band notch design, but bandwidth cannot be controlled. In short, the proposed antenna has improved performance, when compared to all components of other antennas.

Table 1. Performance comparison with previous literature.

Published Literature	Total PCB Size without Height (mm^2)	Bandwidth (GHz)	Notch Band (GHz)	$|S_{11}|$ at Notch (or Notch Quality)	Bandwidth Control/Complex or Simple Structure/ Recon-Figurable	Mutual Coupling (dB)	Gain Var-(dBi)/ Notch Band Gain (dBi)	ECC Using Far-Field Patterns
[26] Anitha et al.	45 × 45	2.2–6.28	No	N.A	N.A	≤ −14	<4/ N.A	<0.25 (uniform)
[17] Asif et al.	50 × 39.8	2.5–12	No	N.A	N.A	≤−17	<N.A/ N.A	N.A
[18] Asif et al.	50 × 39.8	2.7–12	4.8–6.2	−2 dB	Not presented/ complex design structure used to obtain notch/Yes	≤−17	<4/ −3.8	N. A
[14] Keim et al.	60 × 60	2.73–10.68	5.36–6.04	−1.8 dB	Not presented / complex EBG structure/No	<−15	<5/ −3.7	N.A
[16] Wenjing et al.	60 × 60	3.0–16.2	4.0–5.2	−1.4 dB	Not presented / complex EBG structure with vias/No	<−17.5	<6/ −1	<0.3 (uniform)
Proposed antenna	25 × 50	2–12	4.9–6.3	−1 dB	Yes/simple LC stub/ Yes	<−17	<3.5/ −4	<0.15 (uniform) <0.43 (indoor) <0.43 (outdoor)

4. Conclusions

An ultra-compact reconfigurable four-port UWB MIMO antenna is designed and validated after the numerical optimization. PIN diodes are used to reconfigure the WLAN band (4.9–6.3 GHz) by applying a controlled voltage. PIN diodes connect the LC stubs to the ground plane, providing an extra path to current to travel and reject the WLAN band. The rejected bandwidth can be controlled by changing the gap between stub and ground plane. Planar and slanted placement of the radiators allow achieving high compactness. A prototype was fabricated on FR4 laminate; reflection coefficients, mutual coupling, radiation patterns, peak gain, ECC, and TARC were compared with the simulated results showing good agreement. Compactness and reconfigurability of the proposed UWB MIMO antenna make it a very strong candidate for commercial handheld devices.

Author Contributions: M.S.K. provided the idea and performed complete simulations. A.I. performed design, fabrication, and measurements. S.M.A. post processed simulation and measured results. A.-D.C., R.M.S., B.D.B., and D.E.A. assisted in overall paper management, idea development, and manuscript writing.

Funding: This work was supported in part by the EU H2020 Marie Skłodowska-Curie Individual Fellowship through ViSionRF under Grant 840854 and funded by COMSATS University, Islamabad under COMSATS Research Grant Program (Project#16-63/CGRP/CUI/ISB/18/847).

Conflicts of Interest: The authors declare no conflict of interest.

References

1. Kaiser, T.; Zheng, F.; Dimitrov, E., An Overview of Ultra-Wide-Band Systems with MIMO. *Proc. IEEE* **2009**, *97*, 285–312. [CrossRef]
2. Vaughan, R.G.; Andersen, J.B. Antenna diversity in mobile communications. *IEEE Trans. Veh. Tech.* **1978** *36*, 149–172. [CrossRef]
3. Xu, H.-X.; Wang G.-M.; Qi, M.-Q. A miniaturized triple-band metamaterial antenna with radiation pattern selectivity and polarization diversity. *Prog. Electromagn. Res.* **2013**, *137*, 275–292. [CrossRef]
4. *First Report and Order, Revision of Part 15 of the Commissions Rule Regarding Ultra Wideband Transmission Systems*; Federal Communication Commission: Washington, DC, USA, 2002; pp. 2–48.
5. Anagnostou, D.E.; Chryssomallis, M.T.; Braaten, B.D.; Ebel, J.L.; Sepulveda, N. Reconfigurable UWB antenna with RF-MEMS for on demand WLAN Rejection. *IEEE Trans. Antennas Propag.* **2014**, *62*, 602–608. [CrossRef]
6. Zhu, F.; Gao, S.; Ho, A.T.; Abd-Alhameed, R.A.; See, C.H.; Brown, T.W.C.; Li, J.; Wei, G.; Xu, J. Multiple band-notched UWB antenna with band rejected elements integrated in the feed line. *IEEE Trans. Antennas Propag.* **2014**, *61*, 3952–3960. [CrossRef]
7. Li, J.-F.; Chu, Q.-X.; Li, Z.-H.; Xia, X.-X. Compact dual band-notched UWB MIMO antenna with high isolation. *IEEE Trans. Antennas Propag.* **2012**, *61*, 4759–4766. [CrossRef]
8. Zhao, H.; Zhang, F.; Zhang, X.; Wang, C. A compact band-notched ultra-wideband Spatial diversity antenna. *Prog. Electromagn. Res.* **2014**, *51*, 19–26. [CrossRef]
9. Najam, A.; Duroc, Y.; Tedjni, S. UWB-MIMO antenna with novel stub structure. *Prog. Electromagn. Res.* **2011**, *19*, 245–257. [CrossRef]
10. Lee, J.-M.; Kim, K.-B.; Ryu, H.-K.; Woo, J.-M. A compact ultrawideband MIMO antenna with WLAN band-rejected operation for mobile devices. *IEEE Antennas Wirel. Propag. Lett.* **2012**, *11*, 990–993.
11. Yoon, H.K.; Yoon, Y.J.; Kim, H.; Lee, C.H. Flexible ultrawideband polarization diversity antenna with band-notch function. *IET Microw. Antennas Propag.* **2011**, *5*, 1463–1470. [CrossRef]
12. Gao, P.; He, S.; Wei, X.; Xu, Z.; Wang, N.; Zheng, Y. Compact printed UWB diversity slot antenna with 5.5 GHz band-notched characteristics. *IEEE Antennas Wirel. Propag. Lett.* **2014**, *13*, 376–379. [CrossRef]
13. Chacko, B.P.; Augustin, G.; Denidni, T.A. Uniplanar slot antenna for ultrawideband polarization-diversity applications. *IEEE Antennas Wirel. Propag. Lett.* **2013**, *12*, 88–91. [CrossRef]
14. Kiem, N.K.; Phuong, H.N.B.; Chein, D.N. Design of compact 4×4 UWB-MIMO antenna with WLAN band rejection. *Int. J. Antennas Propag.* **2014**, *2014*, 1–11. [CrossRef]
15. Huang, H.; Liu, Y.; Zhang, S.S.; Gong, S.X. Compact polarization diversity ultrawideband MIMO antenna with triple band-notched characteristics. *Microw. Opt. Technol. Lett.* **2015**, *57*, 946–953. [CrossRef]

16. Wu, W.; Yuan, B.; Wu, A. A quad-element UWB-MIMO antenna with band-notch and reduced mutual coupling based on EBG structures. *Int. J. Antennas Propag.* **2018**, *2018*, 1–10. [CrossRef]
17. Khan, M.S.; Capobianco, A.-D.; Asif, S.; Iftikhar, A.; Braaten, B.D. A 4 element compact ultra-wideband MIMO antenna array. In Proceedings of the 2015 IEEE International Symposium on Antennas and Propagation, Vancouver, BC, Canada, 19–25 July 2015.
18. Khan, M.S.; Iftikhar, A.; Asif, S.; Capobianco, A.-D.; Braaten, B.D. A Compact four Elements UWB MIMO antenna with on-demand WLAN rejection. *Microw. Opt. Tech. Lett.* **2016**, *58*, 270–276. [CrossRef]
19. Jung, C.W.; Lee, M.; Li, G.P.; De Flaviis, F. Reconfigurable scanbeam single-arm spiral antenna integrated with RF-MEMS switches. *IEEE Trans. Antennas Propag.* **2006**, *54*, 455–463. [CrossRef]
20. Khan, M.S.; Asif, S.; Capobianco, A.D.; Ijaz, B.; Anagnostou, D.E.; Braaten, B.D. Frequency reconfigurable self-adapting conformal array for changing wedge- and cylindrical-shaped surfaces. *IET Microw. Antennas Propag.* **2016**, *10*, 897–901. [CrossRef]
21. Shelley, S.; Costantine, J.; Christodoulou, C.G.; Anagnostou, D.E.; Lyke, J.C. FPGA-controlled switch-reconfigured antenna. *IEEE Antennas Wirel. Propag. Lett.* **2010**, *9*, 355–358. [CrossRef]
22. Zhao, X.; Riaz, S.; Geng, S. A reconfigurable MIMO/UWB MIMO antenna for cognitive radio applications. *IEEE Access* **2019**, *7*, 46739–46747. [CrossRef]
23. Fatima, A.; Saleem, R; Shabbir, T.; Rehman, S.-U.; Bilal, M.; Shafique, M.F. A compact quad-element UWB-MIMO antenna system with parasitic decoupling mechanism. *Appl. Sci.* **2019**, *9*, 2371.
24. Kumar, P.; Urooj, S.; Alrowais, F. Design of quad-port MIMO/Diversity antenna with tripple-band elimination characteristics for super wideband applications. *Sensors* **2020**, *20*, 624. [CrossRef] [PubMed]
25. Knudsen, M.B.; Pedersen, G.F. Spherical outdoor to indoor power spectrum model at the mobile terminal. *IEEE J. Sel. Areas Commun.* **2002**, *20*, 1156–1168. [CrossRef]
26. Anitha, R.; Vinesh, P.V.; Prakash, K.C.; Mohanan, P.; Vasudevan, K. A compact quad element slotted ground wideband antenna for MIMO applications. *IEEE Trans. Antennas Propag.* **2016**, *64*, 4550–4553. [CrossRef]

Article

Optimal Design of Aperiodic Reconfigurable Antenna Array Suitable for Broadcasting Applications

Ioannis P. Gravas [1,*], Zaharias D. Zaharis [1], Pavlos I. Lazaridis [2], Traianos V. Yioultsis [1], Nikolaos V. Kantartzis [1], Christos S. Antonopoulos [1], Ioannis P. Chochliouros [3] and Thomas D. Xenos [1]

[1] Department of Electrical and Computer Engineering, Aristotle University of Thessaloniki, 54124 Thessaloniki, Greece; zaharis@auth.gr (Z.D.Z.); traianos@auth.gr (T.V.Y.); kant@auth.gr (N.V.K.); chanto@auth.gr (C.S.A.); tdxenos@auth.gr (T.D.X.)

[2] Department of Engineering and Technology, University of Huddersfield, Huddersfield HD1 3DH, UK; p.lazaridis@hud.ac.uk

[3] Hellenic Telecommunications Organization S.A. Member of the Deutsche Telekom Group of Companies, 15122 Athens, Greece; ichochliouros@oteresearch.gr

* Correspondence: igravas@auth.gr

Received: 15 April 2020; Accepted: 13 May 2020; Published: 16 May 2020

Abstract: An aperiodic reconfigurable microstrip antenna array is designed to serve as a DVB-T base station antenna operating in a single broadcasting channel. The antenna array is optimized at 698 MHz (center frequency of DVB-T channel 49) to concurrently achieve a particular radiation pattern shaping with high forward gain, main lobe tilting and null filling inside the service area, as well as low sidelobe level outside the service area, and low standing wave ratio at the inputs of all the array elements. To concurrently satisfy all the above requirements, both the geometry dimensions and the array feeding weights (amplitudes and phases) are optimized, thus leading to a complex multi-variable and multi-objective problem. The problem is solved by applying a recently developed particle swarm optimization (PSO) improved variant, called PSO with velocity mutation, in conjunction with the CST software package, which is employed by the PSOvm every time a full-wave analysis is required. Furthermore, all the optimization methods found in the CST environment are compared with the PSOvm. The results show that the PSOvm is capable of producing an antenna array geometry, which is closer to the predefined requirements than the geometries derived by the rest of the optimizers, in the least amount of computational time.

Keywords: antenna array; aperiodic array; full wave analysis; particle swarm optimization; reconfigurable antenna arrays

---

1. Introduction

Microstrip antenna arrays are very popular as they are low cost and easily fabricated antennas, and therefore can be used as base station antennas for broadcasting. Base station antennas must satisfy multiple requirements, which are essential in broadcasting applications [1–6]. Such requirements are: (i) high forward gain (FG) in order to enhance reception at long distances, (ii) main lobe tilting, as the base station is usually located at places of higher altitude than the service area, (iii) null filling inside an angular sector under the main lobe, as the directional gain must stay above a specific value for proper reception, (iv) low sidelobe level (SLL) so power is not wasted outside the service area, and finally (v) low standing wave ratio (SWR) so power is not wasted inside the feeding network. In this paper, an eight-element aperiodic microstrip antenna array (AMAA) has been designed to satisfy all the above requirements at 698 MHz (center frequency of DVB-T channel 49) and more specifically to exhibit the following characteristics:

1. maximum possible FG at $\theta = 93°$ (θ is the elevation angle),
2. null level ≥ -20 dB with respect to the main lobe peak inside the service area defined from $\theta = 93°$ to $\theta = 143°$ (a 50° service area),
3. $SLL \leq -20$ dB with respect to the main lobe peak outside the service area,
4. $SWR \leq 1.2$ (S-parameters ≤ -21 dB) at the inputs of all the array elements.

The first three requirements imply a significant degradation in the directional gain when passing from the service area to the area where no service is provided. Due to this rapid gain degradation, a transition area is necessary between the previous two areas. Therefore, two transition areas, from $\theta = 80°$ to $\theta = 93°$ and from $\theta = 143°$ to $\theta = 150°$, are defined where the gain will transition from high to low values. Consequently the 3rd requirement must be defined in two angular sectors, the first one ranging from $\theta = 0°$ to $\theta = 80°$ (sector above the main lobe) and the second one from $\theta = 150°$ to $\theta = 360°$. The fourth requirement also helps at enhancing the bandwidth of the antenna by requesting a very low SWR value at the channel center frequency. To satisfy all the above requirements, an AMAA with proper geometry dimensions must be found together with eight appropriate feeding weights (amplitudes and phases). The main lobe tilting is achieved by calculating the proper feeding weights and not mechanically. This is important as the array has big height, since the design frequency is 698 MHz, and it alleviates the mechanical pressure that stresses the whole structure. It is evident that this is a complex multi-variable and multi-objective problem, and such problems can be efficiently solved by using evolutionary optimization algorithms. Furthermore, the antenna studied in this paper can be turned to a reconfigurable one by implementing on-off switches with pin diodes in the feeding network in order to create different feeding weight sets and thus produce different shapes of radiation patterns that provide the desired values of gain, main lobe tilting, null filling and SLL.

Many papers can be found in the literature where different optimization algorithms have been successfully employed to optimize antenna arrays. In Reference [7] three different PSO variants, real-number PSO, binary PSO and multi-objective PSO, are applied to different designs to obtain an optimal solution. The genetic algorithm has been employed in Reference [8] to optimize a quadrifilar helical-spiral antenna for GPS application. In Reference [9] quasi-Newton optimization has been used to design optimal implantable antennas for medical telemetry. Evolutionary techniques combined with machine learning can be found in Reference [10]. The firefly algorithm and particle swarm optimization (PSO) have been employed in Reference [11] to design a uniplanar printed triple band-rejected ultra-wideband antenna. An improved PSO algorithm has been used in Reference [12] to design an aperiodic antenna array with low SLL, minimum half power beamwidth (HPBW) and nulls in desired directions. Finally, the state-of-the-art parallel surrogate model-assisted evolutionary algorithm has been employed in References [13,14]. Since References [7,12] concern aperiodic arrays and are relevant to our paper additional discussion will be made. In Reference [7], one of the optimizations presented, concerns the design of a low SLL 8-element aperiodic array. Only 4 parameters are optimized, specifically the spacing between the array elements and the optimization has 2 objectives, low SLL and low null-to-null beamwidth. The radiation pattern is calculated on a theoretical basis by expressing it as the product of the array factor and the element pattern without taking into account the mutual coupling. However, the element spacing values are allowed to take large values (greater than two times the wavelength) in order to mitigate the mutual coupling. The simulated and fabricated antenna show an $SLL < -10$ dB but it is only shown for $\theta = 0°$ to $\theta = 180°$, thus, without information on the backlobes. There is also no information about the SWR. In Reference [12], as mentioned earlier, an improved PSO variant is used to achieve low SLL, minimum HPBW and placement of nulls in desired directions. The optimization parameters consist of the element spacings and the feeding weights (only amplitudes). The radiation pattern is expressed as the array factor, thus, not taking into account the element pattern and the mutual coupling making the results very theoretical. The radiation patterns are shown for $\theta = 0°$ to $\theta = 180°$, and therefore, it is assumed that the SLL was only optimized in this sector. Also, there is no information on the SWR.

In this paper, the preferred optimizer is an improved (PSO) [15] algorithm that induces mutation to the particles' velocities that did not improve their fitness value in the previous iteration (PSOvm) [16]. The idea behind this method is to help the particles that failed to improve their fitness value in the previous iteration, decouple from their previous velocity which actually did not lead to a fitness improvement. Thus, a perturbation is applied on their previous velocity and by doing so it is expected to help the swarm achieve better exploration and escape when being trapped to a local optimum position. The method has been proved to be very competitive against other well known optimization methods [16] and can handle effectively multi-variable and multi-objective problems as is the case of this paper (the problem consists of 26 optimization parameters and 13 objectives which will be demonstrated in later sections). PSOvm has been implemented in Matlab environment [17]. The AMAA has been fully modelled and simulated in CST Studio Suite (CST from now on) and therefore, the mutual coupling and the element pattern of each array element is considered since full wave analysis is performed. This is really important as it makes the results quite realistic. For each fitness evaluation the antenna array model is updated and a full wave analysis is performed by applying the time domain solver of CST. This solver uses the finite integration technique [18] to solve the electromagnetic problem. Therefore, at each fitness evaluation, the AMAA geometry parameters are updated and the solver simulates the updated model by calling CST built-in visual basic scripts inside the Matlab environment. It is evident that each fitness evaluation is costly in terms of resources and time but the results are highly accurate. In Figure 1 the flowchart of the optimization procedure is depicted. The computer used for the simulations was equipped with an Intel i7 5960X (eightcore) CPU, Nvidia quadro GP100 GPU (suitable for CST GPU-acceleration) with 64 GB DDR4 memory. The CST models employed by the optimization procedure consist approximately of 250,000 mesh cells (as the model changes its dimensions at each fitness evaluation, the amount of mesh cells slightly changes as well), with an average computational time equal to one minute per fitness evaluation. To the best of the authors' knowledge, evolutionary algorithms have never been applied so far to design an AMMA for broadcasting applications in conjunction with CST.

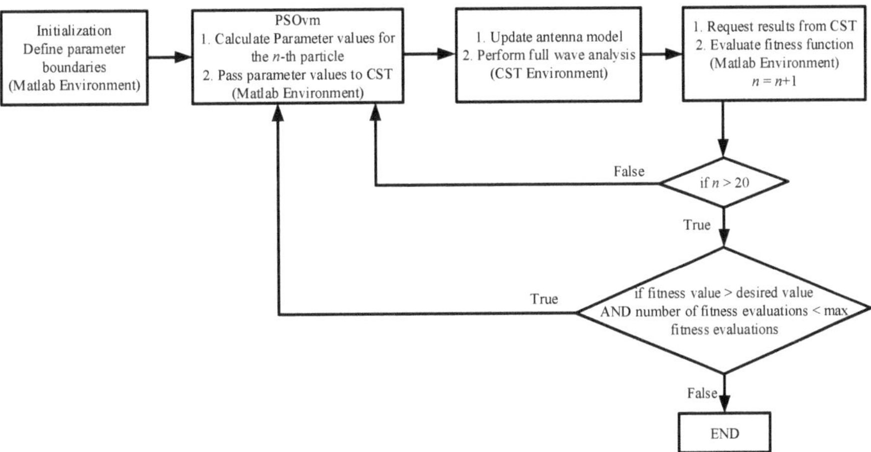

Figure 1. Flowchart of the optimization procedure.

2. Geometry Definition

As mentioned before, CST is employed to model and simulate an eight-element AMAA for DVB-T broadcasting at 698 MHz. The feeding network is out of scope for this paper and therefore the array elements are considered to be directly excited at their input ports by sources. The inset feeding method is used in order to achieve better impedance matching to 50 Ω sources. The geometry of AMAA is illustrated in Figure 2 and is defined by the physical dimensions W and L of the rectangular array elements, four distances d_0, d_1, d_2, d_3 between the array elements, where $d_0 < d_1 < d_2 < d_3$ (aperiodic structure), the transmission line length t_l and width t_w which connects each array element to its respective source, the inset depth s and the inset width g. PSOvm is employed to find the optimal values for all the above parameters. In order to help the optimization algorithm converge faster, the search space for all the geometry parameters must be confined between an upper and a lower limit. As indicated by Reference [19] the theoretical values of W and L are given by:

$$W_{th} = \frac{c}{2f_r} \sqrt{\frac{2}{\epsilon_r + 1}} \tag{1}$$

$$L_{th} = \frac{c}{2f_r \sqrt{\epsilon_{r,eff}}} - 2\Delta L, \tag{2}$$

where c is the free space wave velocity, f_r is the operation frequency (698 MHz in this paper), ϵ_r is the dielectric constant of the substrate used here, $\epsilon_{r,eff}$ is the effective dielectric constant, and ΔL is the length reduction applied at both sides of the effective length to get physical length L. It is noted that the subscript "th" denotes the theoretical value of a variable. The values of $\epsilon_{r,eff}$ and ΔL are respectively estimated [19] by:

$$\epsilon_{r,eff} = \frac{\epsilon_r + 1}{2} + \frac{\epsilon_r - 1}{2} \left[1 + 12\frac{h}{W_{th}}\right]^{-1/2}, \frac{W_{th}}{h} > 1, \tag{3}$$

$$\Delta L = 0.412h \frac{(\epsilon_{r,eff} + 0.3)(W_{th}/h + 0.264)}{(\epsilon_{r,eff} - 0.258)(W_{th}/h + 0.8)}, \tag{4}$$

where h is the substrate height (or thickness). The substrate used in this paper is Duroid RT5880LZ with $\epsilon_r = 2$ and $h = 2.54$ mm [20]. The thickness of the copper cladding (used in CST modeling) at both sides of the substrate is equal to 35 μm [20]. By applying the above equations, we get $W_{th} = 175$ mm and $L_{th} = 151$ mm. It is expected that the optimal values for W and L will not deviate more than 40% from the theoretical values given by (1) and (2) and thus, W is limited between 0.6 W_{th} and 1.4 W_{th} while L is limited between 0.6 L_{th} and 1.4 L_{th}. The distance d_0 is confined between 0.5 λ and 0.8 λ (λ is the free space wavelength at 698 MHz) where the maximum FG is expected to be found. Since d_0, d_1, d_2 and d_3 are in incremental order to create the aperiodical structure of the array, we adopt the rule

$$d_i = d_{i-1} + r_i, \quad i = 1, 2, 3, \tag{5}$$

where r_i, $i = 1, 2, 3$, are random numbers between 0 and 0.2 λ. Therefore, the parameters d_1, d_2 and d_3 are indirectly optimized through r_1, r_2 and r_3, respectively. The values of t_l are confined between 0 and $\lambda/4$ which helps for impedance matching. As for the width of the transmission line t_w, the microwave theory predicts that a width of 8.4 mm will result in a transmission line of characteristic impedance equal to 50 Ω however due to the mutual coupling between the array elements a more relaxed constraint needs to be applied in this variable and therefore the lower limit has been set at half the theoretical value (4.2 mm) and double (16.8 mm) for the upper limit. In order to calculate the theoretical value of d we use the following equations [19]:

$$G_1 = \frac{1}{120\pi^2} \int_0^\pi \left[\frac{\sin\left(\frac{k_0 W_{th}}{2} \cos\theta\right)}{\cos\theta} \right]^2 \sin^3\theta \, d\theta, \tag{6}$$

$$G_{12} = \frac{1}{120\pi^2} \int_0^\pi \left[\frac{\sin\left(\frac{k_0 W_{th}}{2} \cos\theta\right)}{\cos\theta} \right]^2 J_0(k_0 L_{th} \sin\theta) \sin^3\theta \, d\theta, \tag{7}$$

$$R_{in} = \frac{1}{2(G_1 \pm G_{12})} \cos^2 \left(\frac{\pi}{L_{th}} s_{th}\right),$$ (8)

where G_1 is the self conductance of one of the radiating slots of a microstrip patch, k_0 is the free space wavenumber, G_{12} is the mutual conductance between the two radiating slots of a microstrip patch, J_0 is the Bessel function of the first kind of order zero and R_{in} is the resonant input resistance of a patch. By solving (8) for $R_{in} = 50\ \Omega$ we obtain the theoretical value $s_{th} = 44.2$ mm and therefore s is considered to be restricted between 0.6 s_{th} and 1.4 s_{th}. As mentioned earlier, eight feeding weights must be found in order to achieve the desired radiation pattern. The boundaries for the amplitudes of the feeding weights A_n ($n = 1, 2, ..., 8$) are set between 0.1 and 1 (considering an upper limit equal to ten times the lower limit), while the phases P_n are considered to be between $0°$ and $360°$. The optimal value for g according to Reference [21] is considered to be between $W/40$ and $W/10$. As W takes different values during the optimization and g relies on W, it was decided to confine g between $1/40$ and $1/10$ of the current value of W (W_{cur}) during the optimization process. All the boundaries for the optimization parameters are summarized in Table 1. In total, there are 26 parameters to be optimized ($Q = 26$), thus creating a complex problem to solve.

Table 1. Optimization Parameter Limits.

Parameter	Lower Limit	Upper Limit	Parameter	Lower Limit	Upper Limit
W	105 mm	245 mm	t_w	4.2 mm	16.8 mm
L	91 mm	211 mm	s	26.5 mm	61.9 mm
d_0	214 mm	343 mm	g	$W_{cur}/40$	$W_{cur}/10$
r_i, $i = 1,2,3$	0 mm	86 mm	A_n, $n = 1,...,8$	0.1	1
t_l	0 mm	107 mm	P_n, $n = 1,...,8$	$0°$	$360°$

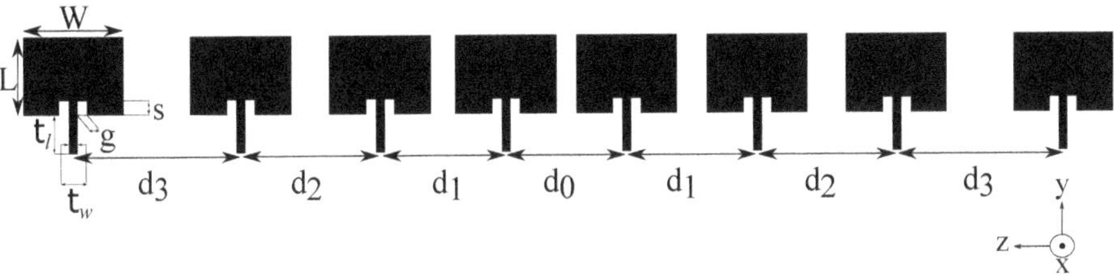

Figure 2. Geometry of the proposed aperiodic microstrip antenna array (AMAA).

3. PSO with Velocity Mutation

There are many PSO variants introduced so far in order to improve the original one. In this variant (PSOvm), the main idea is to apply perturbation to the velocities of the particles that did not improve their fitness value in the previous iteration. In this way, the swarm is protected from premature convergence and achieves greater exploration. PSOvm is based on the *constriction coefficient* PSO (CCPSO) and uses the *gbest* model. In the *gbest* model, one of the components used to update the velocity of a particle is influenced by the best position found by the whole swarm. The other terms are the best position found so far by the particle and its velocity at the previous iteration. By denoting the swarm size as N in a problem with Q optimization variables (Q-dimensional search space), the equations that update the velocity and position of the nth particle ($n = 1, ..., N$), when it improves its fitness at the ith iteration, can be written as:

$$v_{nq}(i+1) = k\{v_{nq}(i) + \phi_1 R[p_{nq}(i) - x_{nq}(i)] + \phi_2 R[g_q(i) - x_{nq}(i)]\}$$ (9)

$$x_{nq}(i+1) = x_{nq}(i) + v_{nq}(i+1),$$ (10)

where v_{nq} and x_{nq} are, respectively, the qth velocity component and the qth position coordinate ($q = 1, ..., Q$) of the nth particle, p_{nq} and g_q are the qth coordinates of the best positions found at the end of the ith iteration by the nth particle and the whole swarm, respectively, R represents random numbers uniformly distributed within the interval $(0, 1)$, ϕ_1 and ϕ_2 are, respectively, the *social* and the *cognitive* coefficient, and are both equal to 2.05, and finally $k = 0.73$ is the *constriction coefficient*. The above equations are the original ones used in CCPSO. However, in PSOvm, if a particle fails to improve its fitness at the previous iteration, then its velocity is updated by using:

$$v_{nq}(i+1) = k\{(0.6 + 0.1m)(2R - 1)v_{nq}(i) + \phi_1 R[p_{nq}(i) - x_{nq}(i)] \\ + \phi_2 R[g_d(i) - x_{nq}(i)]\}, \quad m = 1, ..., 6,$$

(11)

where m is the number of consecutive iterations with no fitness improvement for the nth particle. Practically, the previous velocity of a particle is multiplied by random numbers uniformly distributed in the interval $(-0.7, +0.7)$ when $m = 1$. The boundaries of this interval increase by 0.1 for each consecutive fitness improvement failure. If the nth particle does not improve its fitness six consecutive times, then its velocity is updated by applying (9) and the mutation process stops. This also happens when the particle manages to improve its fitness. The term $(0.6 + 0.1m)(2R - 1)$ as well as the maximum value of m were found after testing PSOvm with several well known multi-dimensional mathematical fitness functions [16].

4. Optimization Results

As mentioned earlier, CST in conjunction with PSOvm is used to find the optimal values for all the parameters specified in Section 2 and Table 1. In this paper, PSOvm employs swarms of 20 particles ($N = 20$). This size was selected after several optimization trials and has shown promising results. As in every optimization problem, a fitness function, usually to be minimized, must be defined in order to guide the optimization algorithm towards the desired results. The expression of the fitness function used here is as follows:

$$Fit = -FG - min(FG - 20, G_{min}) + max(SLL_1, -20) + max(SLL_2, -20) \\ + \sum_{n=1}^{8} max(SWR_n, 1.2),$$

(12)

where FG is the array gain (in dBi) at $\theta = 93°$ which ensures gain maximization and the desired main lobe tilting by $3°$, G_{min} is the minimum gain inside the service area (from $\theta = 93°$ to $\theta = 143°$), SLL_1 represents the SLL inside the angular sector from $\theta = 0°$ to $\theta = 80°$, SLL_2 is the SLL inside the sector defined from $\theta = 150°$ to $\theta = 360°$ and finally SWR_n represents the SWR at the input port of the nth element. The 2nd term in the fitness function ensures that the desired null level of -20 dB is achieved inside the service area, while the 3rd and the 4th terms set the SLL goal of -20 dB for both the angular sectors outside the service area. Finally, the 5th term ensures that all the array elements exhibit an $SWR \leq 1.2$ at their inputs (on the transmission lines edges). The angular sectors from $\theta = 80°$ to $\theta = 93°$ and from $\theta = 143°$ to $\theta = 150°$ serve as transition areas and therefore the radiation pattern inside those areas is not taken into account. The use of *min* and *max* functions in (12) is to help the optimization algorithm in satisfying all the requirements by ignoring any improvement in G_{min}, SLL_1, SLL_2 and SWR_n when the desired values of them are reached. Finally, it must be stressed that the optimization algorithm is tasked to achieve 13 objectives in total, FG maximization and main lobe tilting by $3°$, null level -20 dB, SLL reduction in two sectors SLL_1 and SLL_2, and 8 SWR values by optimizing 26 parameters (described in Section 2) thus, solving a highly complex multi-objective and multi-variable problem.

The dimensions and the feeding weights of the optimized structure are shown in Tables 2 and 3, respectively. Table 4 shows the electromagnetic characteristics of the optimized antenna in the central frequency 698 MHz and Table 5 the input impedances of the array elements. It can be seen that all the requirements have been satisfied. A high FG of 13.9 dBi, main lobe tilting by $3°$ and null level of

−20 dB have been achieved inside the service area. The SLL requirement inside the angular sector from $\theta = 0°$ to $\theta = 80°$ is satisfied achieving −20.8 dB, as well as, the SLL value inside the second angular sector from $\theta = 150°$ to $\theta = 360°$ achieving an SLL of −21 dB. Finally, the SWR for all the array elements is below or equal to 1.2 (S-parameter $\leq$ −21 dB) making the SWR optimization completely successful. Consequently, PSOvm is a capable optimization algorithm and antennas subject to many requirements can be designed. The normalized radiation pattern of the optimized structure is illustrated in Figure 3. Figure 4 shows the S-parameters of all the array elements. It can be seen that all S-parameters are below −21 dB thus achieving the SWR objectives. The optimized antenna exhibits an axial ratio (AR) greater than 22 dB (linear polarization) inside the service area ($\theta = 93°$ to $\theta = 143°$), however, AR was not an optimization goal and is given for consistency. Finally, to verify the optimization results, the optimized structure has been rerun with a finer mesh of approximately 1,000,000 cells (compared to an average of 250,000 cells during the optimization) and the results are essentially the same.

Table 2. Optimized Structure Dimensions.

Parameter	Value (mm)	Parameter	Value (mm)
W	170.9	d_3	348.7
L	152.1	t_l	103.6
d_0	256.9	t_w	13.7
d_1	263.9	s	37.3
d_2	264.8	g	15.8

Table 3. Optimized Feeding Weights.

Parameter	Value	Parameter	Value
$A_1 \angle P_1$	$0.41 \angle 331.9°$	$A_5 \angle P_5$	$0.53 \angle 355.2°$
$A_2 \angle P_2$	$0.67 \angle 333.1°$	$A_6 \angle P_6$	$0.98 \angle 352.1°$
$A_3 \angle P_3$	$0.61 \angle 348.9°$	$A_7 \angle P_7$	$0.72 \angle 58.5°$
$A_4 \angle P_4$	$0.68 \angle 0.65°$	$A_8 \angle P_8$	$0.15 \angle 358.9°$

Table 4. Optimized structure electromagnetic characteristics.

Parameter	Value	Parameter	Value
FG	13.9 dBi	SWR_3	1.07
Main Lobe Tilting	3.0°	SWR_4	1.05
Null level	−20.0 dB	SWR_5	1.16
SLL_1	−20.8 dB	SWR_6	1.16
SLL_2	−21.0 dB	SWR_7	1.20
SWR_1	1.08	SWR_8	1.20
SWR_2	1.11		

Table 5. Input impedances of array elements.

Parameter	Value (Ω)	Parameter	Value (Ω)
Z_1	$53.50 - j2.14$	Z_5	$43.00 + j0.90$
Z_2	$55.23 - j2.42$	Z_6	$53.30 - j7.10$
Z_3	$46.70 + j0.45$	Z_7	$58.30 + j5.40$
Z_4	$50.40 - j2.70$	Z_8	$41.70 + j1.20$

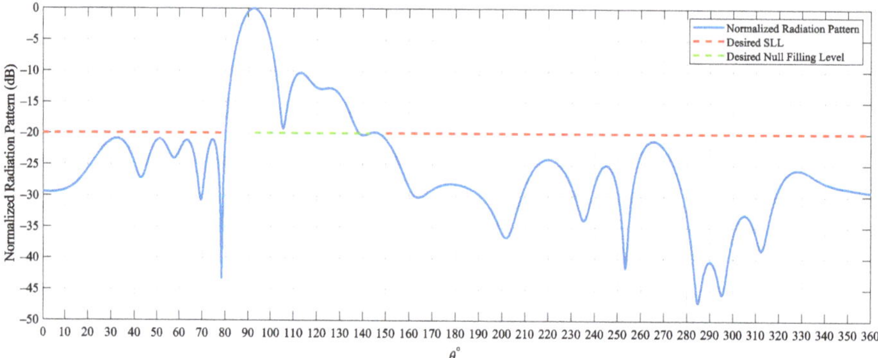

Figure 3. Normalized radiation pattern versus elevation angle θ at azimuth angle $\phi = 0$. The maximum value of the normalized radiation pattern (0 dB) corresponds to the actual maximum gain of the AMAA 13.9 dBi at $\theta = 93°$ (main lobe tilting by 3°). The red dash-line shows the desired SLL outside the service area, and the green dash-line shows the desired null filling level inside the service area. The optimal radiation pattern should be below the red dash-line and above the green dash-line.

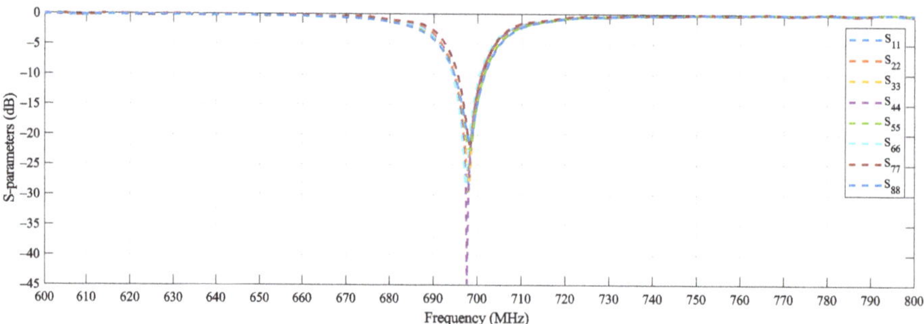

Figure 4. S-parameters of the array elements. All S-parameters are ≤ -21 dB at 698 MHz (DVB-T channel 49 center frequency) satisfying the optimization goals.

5. Comparison with Other Optimization Methods

PSOvm has been compared with all the optimization algorithms that can be found in CST: PSO, genetic algorithm (GA), trusted region framework (TRF), covariance matrix adaptation evolution strategy (CMAES), Nedler Mead simplex algorithm (NMSA), classic Powell (CP) and interpolated quasi Newton (IQN). All the optimization methods have been given a maximum of 2000 fitness evaluations in order to achieve the best result possible. Such benchmarks take a long time to be completed since, as already mentioned in earlier sections, each fitness evaluation is equal to one full wave analysis of the antenna performed by CST with an average computational time of one minute. In Table 6 we show the best result that they achieved within the fitness evaluations limit. In the last column we display the total time (in minutes) required by each method to achieve the best result. PSOvm manages to yield the best result in the least amount of time compared to the other optimization methods with TRF being close. Both PSOvm and TRF converged before hitting the limit of 2000 evaluations. The rest of the optimization methods fail to achieve multiple goals such as main lobe tilting, null level and SLL. The results of this comparison show that both PSOvm and TRF are very capable algorithms that can be used to solve complex multi-variable and multi-objective problems as is the one solved in this paper (26 variables and 13 objectives). Furthermore, it is evident that TRF is the best optimizer available found in CST.

Table 6. Comparison of Optimization Algorithms.

Alg.	Mainlobe Tilting (°)	FG (dBi)	Null Level (dB)	SLL_1 (dB)	SLL_2 (dB)	SWR_1	SWR_2
PSOvm	3.0	13.9	-20.0	-20.8	-21.0	1.08	1.11
PSO	2.2	12.8	-17.8	-6.0	-18.3	1.14	1.08
GA	0.5	10.5	-18.7	-5.7	-15.8	1.06	1.13
TRF	3.0	13.9	-20.5	-20.8	-21.0	1.10	1.12
CMAES	0.5	13.2	-23.4	-13.6	-17.3	1.30	1.21
NMSA	-0.2	12.8	-22.8	-12.6	-12.6	1.22	1.17
CP	-0.1	11.5	-16.6	-8.8	-13.5	1.09	1.15
IQN	-2.0	8.7	-16.8	-7.4	-16.6	1.12	1.13

	SWR_3	SWR_4	SWR_5	SWR_6	SWR_7	SWR_8	Tot. Time (min)
PSOvm	1.07	1.05	1.16	1.16	1.20	1.20	710
PSO	1.09	1.20	1.06	1.26	1.11	1.11	2000
GA	1.16	1.18	1.25	1.12	1.08	1.08	2000
TRF	1.06	1.07	1.16	1.19	1.20	1.20	900
CMAES	1.12	1.20	1.48	1.25	1.15	1.05	2000
NMSA	1.18	1.07	1.05	1.07	1.22	1.06	2000
CP	1.08	1.22	1.27	1.13	1.07	1.12	2000
IQN	1.35	1.19	1.22	1.20	1.32	1.12	2000

6. Conclusions

An aperiodic reconfigurable antenna array that satisfies multiple requirements essential for DVB-T broadcasting, main lobe tilting, high forward gain and null filling inside the service area, low SLL outside the service area and low SWR at the input of all the array elements has been designed. The optimization problem has a significant complexity since 26 parameters have been optimized in order to best satisfy 13 goals. All the goals have been completely satisfied. The results show that the PSOvm is a robust optimization method which can be used to efficiently solve antenna optimization problems of high complexity while saving significant computational time. Also, among the optimizers found in CST, TRF proves to be not only the most efficient of them but also a very competitive one. However, the PSOvm achieves the best result in the least possible computational time compared to those derived by all the other optimizers. Since all the results in this study were obtained by using full wave analysis with CST, they reflect realistic conditions quite well. By applying on-off switches in the feeding network, the antenna array becomes a reconfigurable one, thus providing different shapes of radiation pattern according to the particular requirements that apply to a certain application.

Author Contributions: Conceptualization, Z.D.Z.; Data Curation, I.P.G. and Z.D.Z.; Methodology, Z.D.Z.; Project Administration, Z.D.Z. and T.D.X.; Software, P.I.L., N.V.K. and T.D.X.; Supervision, Z.D.Z., P.I.L., T.V.Y., N.V.K., C.S.A. and T.D.X.; Validation, I.P.G. and Z.D.Z.; Visualization, I.P.G. and Z.D.Z.; Writing—Original Draft, I.P.G.; writing—review and editing, I.P.G., Z.D.Z., P.I.L., T.V.Y., N.V.K., C.S.A., I.P.C. All authors have read and agreed to the published version of the manuscript.

Funding: This research is co-financed by Greece and the European Union [European Social Fund (ESF)] through the Operational Program "Human Resources Development, Education and Lifelong Learning" in the context of the project "Strengthening Human Resources Research Potential via Doctorate Research" under Grant MIS-5000432, implemented by the State Scholarships Foundation (IKY).

Conflicts of Interest: The authors declare no conflict of interest.

References

1. Zaharis, Z.D.; Lazaridis, P.I.; Cosmas, J.; Skeberis, C.; Xenos, T.D. Synthesis of a near-optimal high-gain antenna array with main lobe tilting and null filling using Taguchi initialized invasive weed optimization. *IEEE Trans. Broadcast.* **2014**, *60*, 120–127. [CrossRef]
2. Yamamoto, M.; Arai, H.; Ebine, Y.; Nasuno, M. Simple design of null-fill for linear array. In Proceedings of the International Symposium on Antennas and Propagation, Okinawa, Japan, 24–28 October 2016; pp. 410–411.
3. Zaharis, Z.D. Radiation pattern shaping of a mobile base station antenna array using a particle swarm optimization based technique. *Electr. Eng.* **2008**, *90*, 301–311. [CrossRef]

4. Patidar, H.; Mahanti, G.K.; Muralidharan, R. Synthesis of non-uniformly spaced linear array of unequal length parallel dipole antennas for impedance matching using QPSO. *Int. J. Microw. Opt. Technol.* **2017**, *12*, 172–181.

5. Lazaridis, P.I.; Tziris, E.; Zaharis, Z.D.; Xenos, T.; Holmes, V.; Cosmas, J.P.; Glover, I.A. Comparative study of broadcasting antenna array optimization using evolutionary algorithms. In Proceedings of the 2016 URSI Asia-Pacific Radio Science Conference (URSI AP-RASC), Seoul, Korea, 21–25 August 2016; pp. 1299–1301, doi:10.1109/URSIAP-RASC.2016.7601166. [CrossRef]

6. Zhang, Y.; Zhang, X.Y.; Ye, L.H.; Pan, Y.M. Dual-band base station array using filtering antenna elements for mutual coupling suppression. *IEEE Trans. Antennas Propag.* **2016**, *64*, 3423–3430. [CrossRef]

7. Jin, N.; Rahmat-Samii, Y. Particle swarm optimization for antenna designs in engineering electromagnetics. *J. Artif. Evol. Appl.* **2008**, *2008*, 1–10. [CrossRef]

8. Zhou, D.; Gao, S.; Abd-Alhameed, R.A.; Zhang, C.; Alkhambashi, M.S.; Xu, J.D. Design and optimisation of compact hybrid quadrifilar helical-spiral antenna in GPS applications using Genetic Algorithm. In Proceedings of the 2012 6th European Conference on Antennas and Propagation (EUCAP), Prague, Czech Republic, 26–30 March 2012; pp. 1–4.

9. Kiourti, A.; Nikita, K.S. Accelerated design of optimized implantable antennas for medical telemetry. *IEEE Antennas Wirel. Propag. Lett.* **2012**, *11*, 1655–1658. [CrossRef]

10. Liu, B.; Aliakbarian, H.; Ma, Z.; Vandenbosch, G.A.; Gielen, G.; Excell, P. An efficient method for antenna design optimization based on evolutionary computation and machine learning techniques. *IEEE Trans. Antennas Propag.* **2013**, *62*, 7–18. [CrossRef]

11. Mohammed, H.J.; Abdullah, A.S.; Ali, R.S.; Abd-Alhameed, R.A.; Abdulraheem, Y.I.; Noras, J.M. Design of a uniplanar printed triple band-rejected ultra-wideband antenna using particle swarm optimisation and the firefly algorithm. *IET Microw. Antennas Propag.* **2016**, *10*, 31–37. [CrossRef]

12. Rahman, S.U.; CAO, Q.; Ahmed, M.M.; Khalil, H. Analysis of linear antenna array for minimum side lobe level, half power beamwidth, and nulls control using PSO. *J. Microw. Optoelectron. Electromagn. Appl.* **2017**, *16*, 577–591. [CrossRef]

13. Liu, B.; Koziel, S. Antenna array optimization using surrogate-model aware evolutionary algorithm with local search. In Proceedings of the 2015 IEEE International Symposium on Antennas and Propagation & USNC/URSI National Radio Science Meeting, Vancouver, BC, Canada, 19–24 July 2015; pp. 1330–1331.

14. Liu, B.; Akinsolu, M.O.; Ali, N.; Abd-Alhameed, R. Efficient global optimisation of microwave antennas based on a parallel surrogate model-assisted evolutionary algorithm. *IET Microw. Antennas Propag.* **2019**, *13*, 149–155. [CrossRef]

15. Eberhart, R.; Kennedy, J. A new optimizer using particle swarm theory. In Proceedings of the MHS'95, Sixth International Symposium on Micro Machine and Human Science, Nagoya, Japan, 4–6 October 1995; pp. 39–43.

16. Zaharis, Z.D.; Gravas, I.P.; Yioultsis, T.V.; Lazaridis, P.I.; Glover, I.A.; Skeberis, C.; Xenos, T.D. Exponential Log-Periodic Antenna Design Using Improved Particle Swarm Optimization With Velocity Mutation. *Trans. Magn.* **2017**, *53*, 1–4. [CrossRef]

17. Mathworks Matlab. Available online: https://www.mathworks.com (accessed on 12 May 2018).

18. CST Studio Suite. Available online: https://www.3ds.com/products-services/simulia/products/cst-studio-suite/ (accessed on 16 June 2019).

19. Balanis, C.A. *Antenna Theory: Analysis and Design*; Wiley-Interscience: New York, NY, USA, 2005; pp. 816–826.

20. Rogers Corporation. Available online: https://www.rogerscorp.com (accessed on 2 February 2020).

21. Matin, M.A.; Sayeed, A.I. A design rule for inset-fed rectangular microstrip patch antenna. *WSEAS Trans. Commun.* **2010**, *9*, 63–72.

Article

Metasurfaces for Reconfiguration of Multi-Polarization Antennas and Van Atta Reflector Arrays

Mohammed Alharbi [†], Meshaal A. Alyahya [†], Subramanian Ramalingam, Anuj Y. Modi, Constantine A. Balanis * and Craig R. Birtcher

School of Electrical, Computer, and Energy Engineering, Arizona State University, Tempe, AZ 85287, USA; msalharb@asu.edu (M.A.); malyahy3@asu.edu (M.A.A.); subramanian.ramalingam@asu.edu (S.R.); aymodi1@asu.edu (A.Y.M.); CRAIG.BIRTCHER@asu.edu (C.R.B.)

* Correspondence: balanis@asu.edu
† These authors contributed equally to this work.

Received: 9 July 2020; Accepted: 4 August 2020; Published: 6 August 2020

Abstract: This paper discusses the application of metasurfaces for three different classes of antennas: reconfiguration of surface-wave antenna arrays, realization of high-gain polarization-reconfigurable leaky-wave antennas (LWAs), and performance enhancement of van Atta retrodirective reflectors. The proposed surface-wave antenna is designed by embedding four square ring elements within a metasurface, which improves matching and enhances the gain when compared to conventional square-ring arrays. The design for linear polarization comprises of a 1×4 arrangement of ring elements, with a 0.56λ spacing, placed amidst periodic patches. A 2×2 arrangement of ring elements is utilized for reconfiguration from linear to circular polarization, where a similar peak gain with better port isolation is realized. A prototype of the 2×2 array is fabricated and measured; a good agreement is observed between simulations and measurements. In addition, the concepts of the design of polarization-diverse holographic metasurface LWAs that form a pencil beam in the desired direction with a reconfigurable polarization are discussed. Moreover, recent developments incorporating polarization-reconfigurability in metasurface LWAs are briefly reviewed. In the end, the theory of van Atta arrays is outlined and their monostatic RCS is reviewed. A conventional retrodirective array is designed using aperture-coupled patch antennas with a microstrip-line feeding network, where the scattering from the structure itself degrades the performance of the reflector. This is followed by the integration of judiciously synthesized metasurfaces to reconfigure and improve the performance of retrodirective reflectarrays by removing the above-mentioned undesired scattering from the structure.

Keywords: reconfigurable antennas; surface wave array antenna; metasurface ground plane; low-profile; square-ring antennas; van Atta reflector; RCS reduction; structural mode scattering; metasurfaces; polarization reconfigurability; leaky waves

---

1. Introduction

Metasurfaces are two-dimensional, less bulky, and low-loss equivalents of metamaterials [1]. They have been popular as they are suitable for a variety of applications that include wavefront shaping and control [2,3], beamforming using high-gain leaky-wave antennas (LWAs) [4–6], radar cross section (RCS) reduction [7–9], and design of low-profile antennas [10,11]. These metasurfaces are realized as an array of patches on a grounded dielectric substrate, and they are easy to fabricate. By changing the geometry of these patches, their surface impedance and reflection phase can be controlled. A zero reflection phase can be realized at a particular frequency, thus mimicking a perfect magnetic conductor (PMC). Hence, they have been utilized to enhance the performance of antennas, in addition to the design of reconfigurable antennas. This paper details the utilization of metasurfaces to design and enhance the performance of three types of antennas: surface-wave antenna arrays, polarization-reconfigurable holographic LWAs, and van Atta retrodirective reflectors.

1.1. Surface-Wave Antenna Arrays

Proper excitation of surface waves enhances the overall antenna performance [12,13]. Consequently, several printed antennas have been reported, where the influence of surface waves has been examined. Monopole-like radiation has been achieved using a circular patch in the vicinity of uniform (square) and non-uniform metasurfaces [14,15]. Similarly, surface waves have been excited utilizing a square ring element surrounded by anisotropic metasurfaces, thus yielding broadside radiation [11]. Another design achieving broadside radiation has been realized by placing a diamond-shaped patch in the vicinity of square-patch metasurfaces [16].

To further improve the radiation performance, the analysis has been expanded to include multiple radiators in the vicinities of metasurfaces to form array metasurface antennas [16,17]. The reported designs require a specific spacing between the metasurface supercell to maintain and realize a high broadside gain. This spacing degrades the aperture efficiency and reduces the maximum attained gain. As it will be demonstrated in this paper, the spacings between the metasurface supercell can be reduced without disturbing the functionality of the design. The aforementioned designs are linearly polarized and have a wide range of potential low-profile applications. Circular polarization can be realized by placing a linear-to-circular metasurface superstrate above radiating elements [18]. This approach requires a certain spacing between the radiating element and the metasurface superstrate, which disturbs the overall height-profile. The second approach is to excite surface waves in two orthogonal planes that share the same magnitude and $90°$ phase difference [19–21]. A truncated square patch element positioned in close proximity to square-patch metasurfaces resulted in 7 dB peak gain, a fractional bandwidth of 45.6%, and 3-dB axial ratio (AR) bandwidth of 23.4% [19]. Another design implemented a planar slot to excite rectangular-patches metasurfaces and led to an average gain of 5.8 dB, an impedance bandwidth of 33.7%, and a 3-dB AR bandwidth of 16.5% [20]. The low peak gain observed in the above-mentioned designs is improved to 12 dB by forming a 2×2 array of truncated patch elements positioned below square-patches metasurface [21]. This achieved gain can be further promoted by exciting surface waves along with the active element's fundamental mode, which will be demonstrated.

In the first part of this paper, which corresponds to surface-wave antenna arrays, an array of multiple square ring elements embedded within a metasurface is presented. With different array configurations, linear and circular polarizations are achieved.

1.2. Polarization-Reconfigurable Holographic LWAs

Metasurfaces have facilitated the realization of surface-impedance modulated periodic LWAs that are capable of forming fan or pencil beams with high gains and narrow beamwidths. One-dimensional LWAs form a fan beam in the desired angular direction, whereas two-dimensional LWAs form a pencil beam. The angular direction of the formed beam is determined by the modulation parameters of average surface reactance and period of modulation. One-dimensional periodic metasurface LWAs were realized in [22], while two-dimensional LWAs were designed using the principle of holography in [4].

The polarization of the formed beam is dependent on the polarization of the leaky-wave mode; TM-mode leaky waves result in vertically-polarized beams, while TE-mode leads to horizontally polarized [23]. Circularly-polarized beams were realized by employing a spiral surface impedance modulation in [24]. The design of scalar holographic metasurfaces to form a pencil beam in the desired direction with a desired polarization was detailed in [25]. Tensor impedance surfaces with amplitude, phase, and polarization control were proposed in [26].

The suitability of these metasurfaces to communication-based applications necessitates the formation of beams with reconfigurable polarizations. The design of polarization-diverse metasurfaces that can form a pencil beam in the desired direction with a horizontal, vertical, or circular polarization is discussed in [6]. The polarization state was reconfigured by changing the source of excitation. A similar approach was used to realize dual reconfigurable polarizations using tensor impedance surfaces, and recently proposed in [27]. Another recently-proposed design achieves polarization reconfigurability by using polarization-insensitive holographic surfaces, where the reconfiguration is facilitated by diodes on the surface-wave launchers [28].

The second part of this paper reviews the procedures to design a multi-polarization metasurface LWA. The existing techniques are reviewed and potential new techniques are proposed.

1.3. Van Atta Retrodirective Reflectors

In many radar and communication systems, reflectors are utilized because of their ability to maximize the re-radiation towards the direction of wave incidence. Such a response can be achieved using a retrodirective reflector, first proposed in [29]. The retrodirective reflector is an antenna array whose elements are interconnected by transmission lines such that the received signal is then reradiated towards the direction of incidence [30–32]. In automotive collision avoidance systems, a high scattered field can be achieved only toward near-normal directions to the target surface. Thus, by equipping vehicles and road obstacles with retrodirective reflectors, the self-phasing feature of such reflectors will increase scattering beamwidths and targets will become more visible [32]. Furthermore, retrodirective reflectors have been investigated for the application of wireless power transfer [33–35]. A device that requires wireless power can send a beacon signal that is then received, amplified, and sent back to the user by the retrodirective antenna array.

Generally, retrodirective arrays can be synthesized with basic radiating elements such as dipoles [31]. In addition, they can be designed using patch antennas [36–38]. However, due to the high-backscattering from the structure (consisting of antenna arrays), it degrades the performance of the retrodirective reflector by destructively interfering with the desired reradiated fields. Thus, a low-backscattering array of long slots was investigated [39], and it was shown that this reflector could reradiate fields without interference from the array structure's scattering. However, a retrodirective reflector comprised of a patch antenna array is a better option due to the simplicity of design and the low profile of patch antenna elements if the high backscattering by the patch antenna array can be mitigated. Thus, artificial magnetic conductor (AMC) technology is utilized to reconfigure the performance of such antenna arrays by reducing scattering from such structures [40].

In the third part of this paper, a van Atta retrodirective reflector with a smoother backscattering pattern is synthesized and developed using a two-dimensional microstrip-antenna array. Conventional retrodirective reflectors are sensitive to the interference by the fields scattering from the antenna structure. Using a virtual feeding network, structural mode scattering is identified and canceled using metasurfaces.

2. Surface-Wave Antenna Arrays

This section details the first of the three areas, where metasurfaces are used to realize and enhance the antenna performance. In this section, a 1×4 arrangement of square ring elements is printed on the same plane with periodic patches to achieve linear polarization. With a spacing of $0.56\lambda_0$ between the elements, a high realized gain with high aperture efficiency is attained. The polarization can be reconfigured from linear to circular using a 2×2 structure with a spacing of $0.85\lambda_0$ between the ring elements. After analyzing the port isolations between the ports, a simple parallel feeding network is designed and incorporated with the structures. Due to the similar characteristics of the two arrays, only the 2×2 circular polarized structure is experimentally validated.

2.1. 1 × 4 Linearly Polarized Metasurface Array

The array designs considered in this paper are based on the structure proposed in [11]. The radiator within the metasurface (RWMS) is extended due to its wide matching bandwidth and stable unidirectional radiation. The mechanism of this design depends on the excitation of the radiating elements' mode (TM11) and the surface-wave modes in such a way that their resonances are closely spaced in order to enhance the overall antenna performance in terms of gain and bandwidth. For brevity, this paper does not cover the analysis and the design of RWMS; the readers can refer to [11] to develop an understanding of the RWMS design. A one-dimensional (1-D) array is designed, as illustrated in Figure 1a. The array has four square-ring elements utilized to launch surface waves, in addition to their resonances. The inter-element spacing is $s = 37.5$ mm, which is about $0.56\lambda_o$ at 4.5 GHz. The substrate is Rogers RT/Duroid-5880 ($\varepsilon_r = 2.2$, $\tan\delta = 0.0009$) with a thickness $h = 5.08$ mm. The ring elements and the metasurface patches are printed on the top face of a single substrate with length $l_s = 191.5$ mm and width $w_s = 62$ mm. The remaining parameters are detailed in the caption of Figure 1. Since the design relies on the excitation of surface waves, it would be of interest to shed insight on the isolation between the ports. Therefore, the ring elements are individually fed with four coaxial probes, as presented in Figure 1a. For optimum broadside radiation, all elements are excited with the same magnitude and phase.

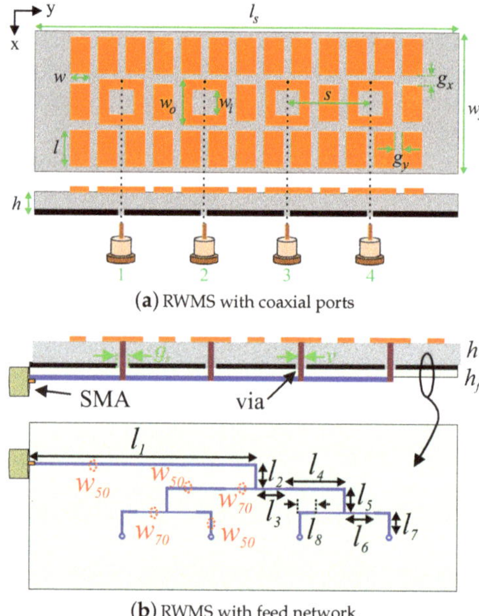

Figure 1. The 1 × 4 square ring elements embedded within metasurface and excited with (**a**) four coaxial probes and (**b**) feed network. The RWMS parameters are: $w_i = 10$ mm, $w_o = 20$ mm, $l = 16$ mm, $w = 9$ mm, $g_x = 4.5$ mm, and $g_y = 3.5$ mm. The feed network parameters are: $l_1 = 96.3$ mm, $l_2 = 10$ mm, $l_3 = 11.85$ mm, $l_4 = 25.65$ mm, $l_5 = 10$ mm, $l_6 = 11.85$ mm, $l_7 = 9.29$ mm, $l_8 = 6.9$ mm, $g_s = 2.74$ mm, $v = 1.143$ mm, $w_{50} = 1.1$ mm, and $w_{70} = 0.7$ mm.

The reflection coefficients and the mutual coupling between the coaxial ports are simulated and presented in Figure 2a. The −10 dB reflection coefficient bandwidth for all ports (dotted red line) is about 28% from 3.9–5.2 GHz, which is the same as attained with a single RWMS [11]. The isolation between two adjacent elements (e.g., S_{12}) is low at lower frequencies and reaches up to −10 dB at 5.2 GHz. This is attributed to surface waves along the y-axis, which begin to resonate above 5 GHz [11], and they disturb both the port isolation and the broadside radiation. The isolation can be improved by increasing the spacing between the square-ring elements; however, this approach will introduce undesired sidelobes.

Once the port isolations are examined, a parallel feeding network, presented in Figure 1b, is designed for the implementation of the RWMS array. The feeding network is also designed on Rogers RT/Duroid-5880 substrate whose thickness $h_f = 0.38$ mm. The feed is stacked below the 1×4 array, and the outputs of the feed are coupled to the ring elements by metallic vias. The dimensions of the feed network are detailed in the caption of the figure. For comparison purposes, the S-parameters investigated for the feed network when it is connected to the RWMS and a conventional square element array (in the absence of metasurface). As illustrated in Figure 2a, when the feed is connected to the RWMS array, the -10 dB fractional bandwidth is comparable to that obtained with coaxial ports; however, poor matching is attained when the feed is integrated with a conventional 1×4 square ring array.

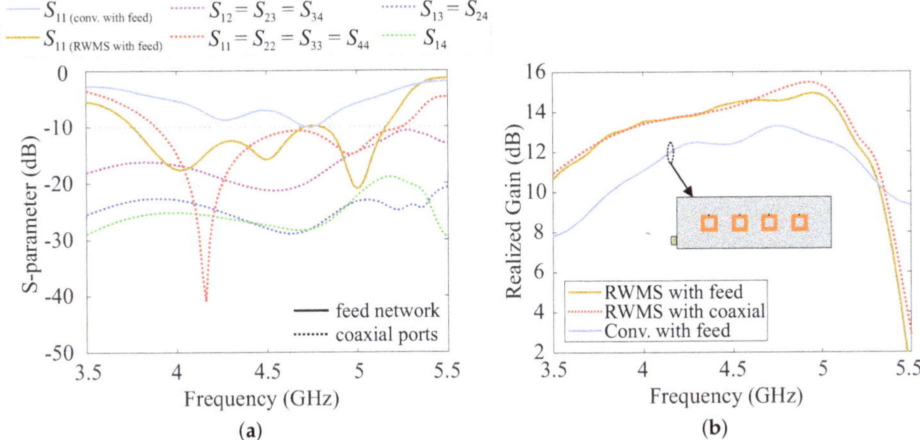

Figure 2. (a) S-parameters and (b) the broadside realized gain of the 1×4 RWMS array.

To further examine the influence of coupling on the radiation behavior, the broadside realized gain is examined and plotted in Figure 2b. At lower frequencies, the gain is the same when the RWMS is excited either by the feed network or the multiple coaxial probes; however, the two results begin to deviate slightly around 4.7 GHz, which is attributed to the more pronounced effect of coupling.

The simulated realized gain, when the RWMS is excited with a feed network, ranges from 13.5 dB at 4 GHz and reaches up to 14.8 dB at 5 GHz. This observed high gain for the RWMS is attributed to the use of metasurfaces, as displayed in Figure 2b. The aperture efficiency at 4.5 GHz is about 83% for the RWMS array, while it is only 52% for a conventional square ring array. The impact of the metasurface in enhancing the efficiency is demonstrated by plotting the surface current distribution at 4.5 GHz in Figure 3. As observed, both the ring elements and the periodic patches contribute to the total radiation; this justifies the higher achieved efficiency compared to the conventional array.

Figure 3. Surface current distribution at 4.5 GHz for the 2 × 2 metasurface array.

2.2. 2 × 2 Circularly Polarized Metasurface Array

This section addresses the reconfiguration from the linear to circularly polarized RWMS array. Since RWMS is designed to excite surface waves on a single plane [11], circular polarization (CP) can be achieved by arranging orthogonal structures of RWMS fed with a 90-degree phase difference. To achieve a gain comparable to the linear RWMS, a 2 × 2 RWMS scheme is considered and designed, as displayed in Figure 4a.

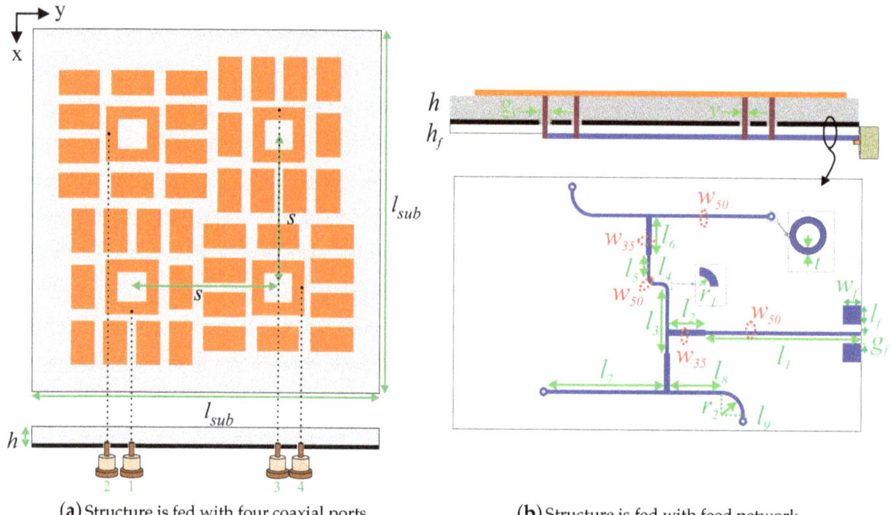

(**a**) Structure is fed with four coaxial ports

(**b**) Structure is fed with feed network

Figure 4. 2 × 2 square ring elements embedded within metasurface and excited with (**a**) four coaxial probes, and (**b**) feed network. The feed network parameters are: $l_1 = 50.875$ mm, $l_2 = 12$ mm, $l_3 = 20$ mm, $l_4 = 2.25$ mm, $l_5 = 7.2$ mm, $l_6 = 11.9$ mm, $l_7 = 37.87$ mm, $l_8 = 16.92$ mm, $l_9 = 0.69$ mm, $w_{50} = 1.1$ mm, $w_{35} = 1.9$ mm, $g_s = 2.74$ mm, $v = 1.143$ mm, $w_f = l_f = 6$ mm, $g_f = 2.45$ mm, $r_1 = 1.4$ mm, $r_2 = 6.5$ mm, and $t = 0.79$ mm.

A square grounded dielectric is modeled with a thickness $h = 5.08$ mm, and a length $l_{sub} = 133$ mm. The spacing s along the x- and y-planes is 56 mm, which corresponds to $0.85\lambda_o$ at 4.5 GHz. The dielectric material, the dimensions of the ring elements, and the metasurface patches are identical to those given in the 1 × 4 array configurations. Similar to the linear array, the coupling between the ring elements is investigated first by individually feeding the ring elements with four coaxial probes. The simulated scattering parameters for 2 × 2 circular polarized structure are depicted in Figure 5a. The −10 dB fractional bandwidth is slightly less than that for the 1 × 4 array, with a bandwidth of 25% from 3.9 to 5.0 GHz. The mutual coupling between vertically/horizontally adjacent ports and diagonally orientated ports is less than −20 dB, which is lower than those observed for the 1 × 4 linear array. This is attributed to the orthogonality of the excited surface waves, which agrees with the findings in [41].

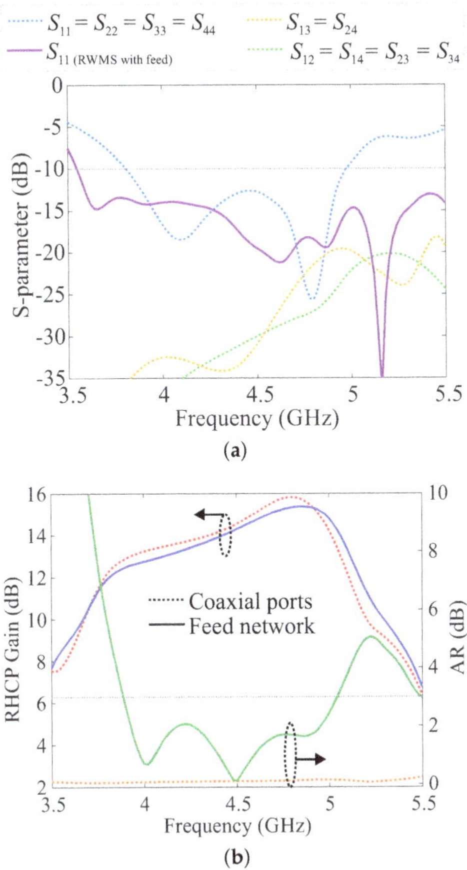

Figure 5. (a) S-parameters and (b) the broadside RHCP gain along with the axial ratio of the 2 × 2 RWMS scheme.

Once the coupling is examined, a parallel feeding network is designed to excite all four ring elements with a single coaxial feed, as illustrated in Figure 4b. The feed is designed at the center frequency (4.5 GHz) to split the power equally among the four ring elements with a 90° phase increment. The feed can dictate whether the sense of rotation of the CP is right hand (RHCP) or left hand (LHCP), depending on the orientation of the phase increment (clockwise or counterclockwise); in this work, the feed is modeled to realize RHCP. The feed is designed on Rogers RT/Duroid-5880 substrate, which has a thickness $h_f = 0.38$ mm. Similar to the 1 × 4 linear array, the feeding substrate is stacked below the 2 × 2 circular array, and the outputs of the feed network are coupled to the ring elements using vias. The remaining dimensional parameters of the feeding network are detailed in the caption of Figure 4. When the feed network excites the 2 × 2 RWMS, the −10 dB matching bandwidth is larger than that obtained with the four coaxial feeds. This is attributed to the parallel feed structures which do not provide any isolation between the ports; therefore, any reflected power from any port can couple through to other ports resulting in small power return back to the feed.

To further analyze the performance of the 2 × 2 RWMS structure, the broadside realized gain and the axial ratio are examined. The simulated results for the two different feed approaches are presented in Figure 5b.

When the structure is fed with individual probes, the realized gain is 13.5 dB at 4 GHz and reaches up to 15.8 dB at 4.8 GHz. Almost similar gain is attained when the feed network is utilized. However, there is a noticeable discrepancy in the axial ratio between the two feeding methods. For the individual feed approach, the broadside axial ratio (AR) is below 0.25 dB over the design frequency range, while it is only below 3 dB over a more narrow frequency range (3.9 GHz to 5.1 GHz) when the structure is excited with the feed network. The fluctuation in the axial ratio magnitude is due to the poor isolation of the considered feeding network, which leads to magnitude and phase errors for the power reached to each port [42].

The broadside realized gain at the center frequency (4.5 GHz) is around 14.5 dB, while it is only 10 dB for a 2 × 2 conventional array. This corresponds to the aperture efficiencies of 56% and 20% for the metasurface array and its conventional counterpart, respectively. Similar to the linear metasurface array, the high attained efficiency for the 2 × 2 metasurface array is due to the excitation of the radiating element's mode along with the surface waves' modes. This can be verified by examining the surface current distribution in Figure 6, which shows that both the radiating elements and the periodic patches contribute to the total radiation.

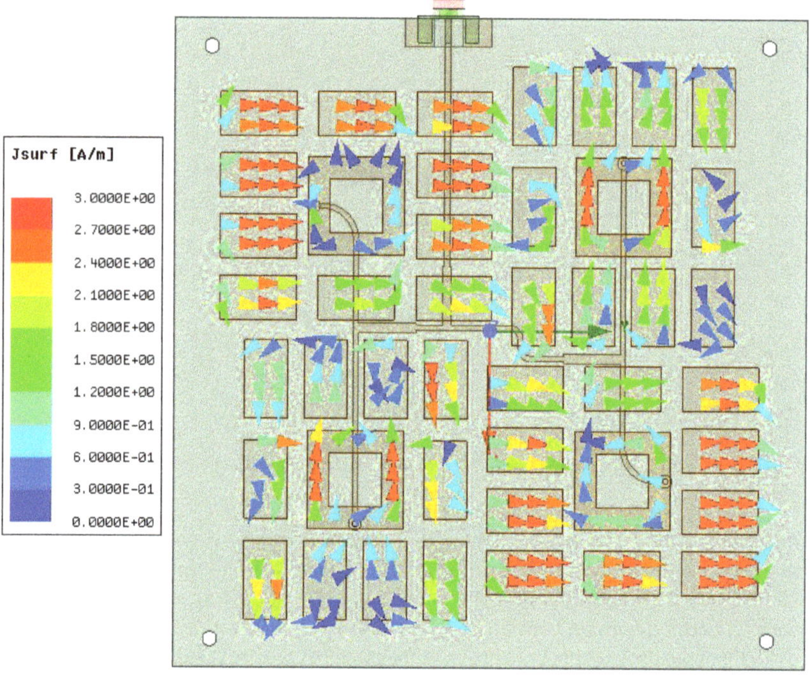

Figure 6. Surface current distribution at 4.5 GHz for the 2 × 2 metasurface array.

2.3. Fabrication and Measurements

To experimentally validate the simulated data, the proposed 2 × 2 RWMS circular array was fabricated, as depicted in Figure 7.

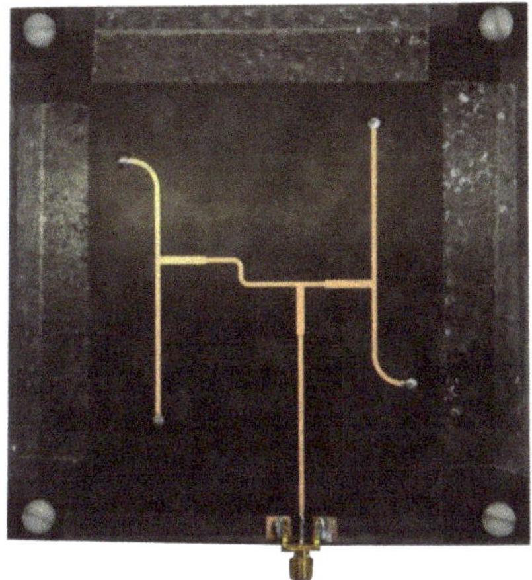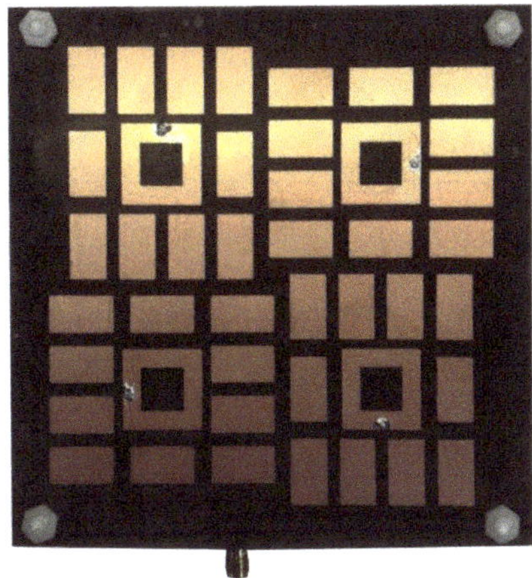

Figure 7. Top and bottom views of the fabricated prototype of the 2 × 2 RWMS array incorporated with the feed network.

Since linear and circular RWMS arrays share similar characteristics, only the circular array was considered for fabrication. The substrate materials and the dimensions of the fabricated design were identical to those specified in Figure 4. An edge mount SMA connector was soldered to the feed network substrate, as demonstrated in Figure 7. Four nylon screws were placed at the corner to tighten the feed and RWMS substrates together and avoid any misalignment.

The comparison of the measured and simulated results are displayed in Figures 8 and 9. The simulated and measured S_{11} and the axial ratio showed a good agreement; however, the broadside RHCP gain was about 1 dB lower than the simulated data. This deviation could be attributed to the fabrication tolerance and the stacking assembly error. The measured and simulated radiation patterns along different planes are displayed in Figure 9; a good agreement was observed between the simulated and measured data.

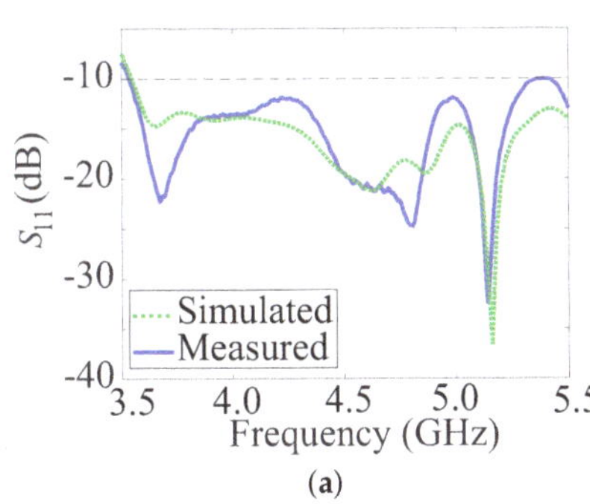

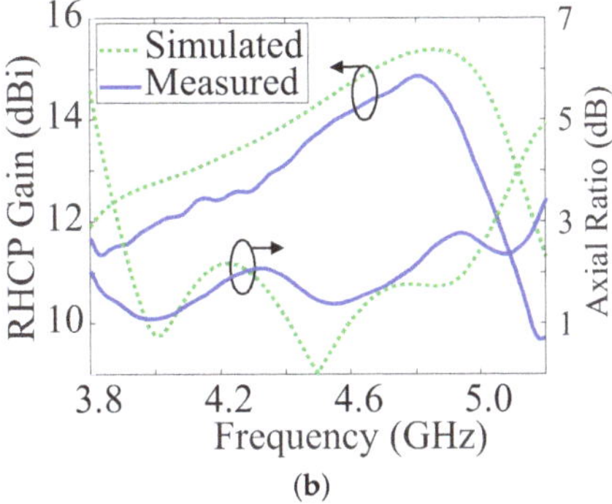

Figure 8. Comparison between simulated and measured values of the (a) reflection coefficient, (b) RHCP broadside gain, and axial ratio.

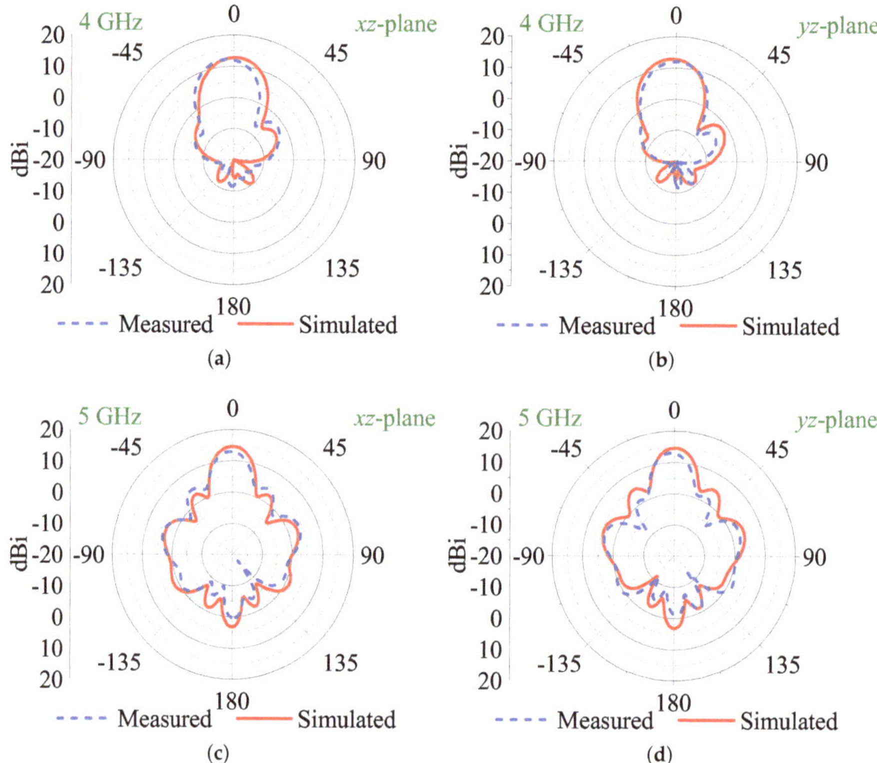

Figure 9. Simulated and measured radiation patterns at (**a**) 4 GHz along *xz*-plane, (**b**) 4 GHz along *yz*-plane, (**c**) 5 GHz along *xz*-plane and (**d**) 5 GHz along *yz*-plane.

3. Polarization Reconfigurable Holographic LWAs

The advent of metasurfaces has facilitated the realization of surface-reactance modulation by varying the geometries of patches that are periodically arranged on a dielectric-covered ground plane. Patches with an inductive reactance support TM-mode surface waves while capacitive surface reactance support TE-mode [43]. The surface reactance of a metasurface unit cell depends on the frequency of operation and can be determined using the transverse-resonance method (TRM) [23].

The surface-reactance modulation function X_s required to form a pencil beam, in the desired elevation angle θ_d (from broadside) and the azimuth angle ϕ_d, is obtained using the holographic principle [4,23]

$$X_s(x,y) = X_{avg} \left\{ 1 + M e^{j\beta\sqrt{(x-x_c)^2+(y-y_c)^2}} e^{jk_0(x\sin\theta_d\cos\phi_d + y\sin\theta_d\sin\phi_d)} \right\} \tag{1}$$

where X_{avg} is the average surface reactance, M is the modulation index, (x_c, y_c) is the location of the source and k_0 is the free-space phase constant. This leads to a holographic pattern which is a set of periodic ellipses that have a focus at the location of the source, as shown in Figure 10.

The metasurface forms a pencil beam with a linear polarization depending on the leaky-wave mode. The pencil beam is formed as a superposition of radiation from forward and backward leaky waves that move in opposite directions with a change in frequency. Hence, they have to be operated at the phase-crossover frequency, where the radiation from the forward and backward leaky waves interfere constructively.

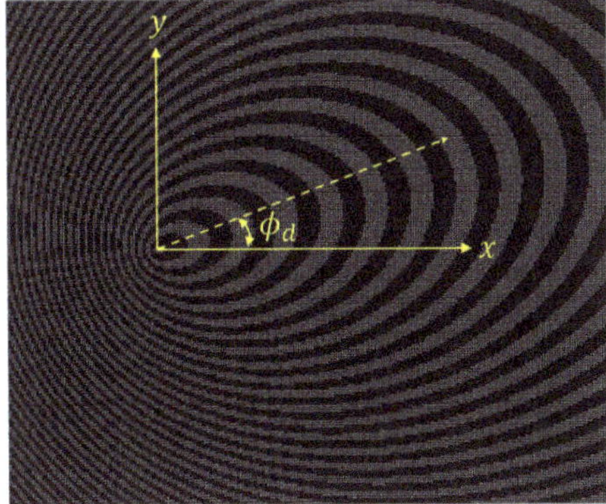

Figure 10. The holographic pattern of the surface-reactance profile needed to form a pencil beam along (θ_d, ϕ_d).

Different polarizations can be obtained by performing geometric modifications along the surface [25]. While operating at non-phase crossover frequencies, vertical and horizontal polarizations can be achieved by including a 180° phase shift on either side of the principal axes of the ellipses; a 180° phase shift along the minor axis results in vertical polarization, while that along the major axis leads to horizontal. A phase shift of 180° on either side of both the major and minor axes, along with a relative phase of 90° between them, leads to a circularly polarized pencil beam. The principal axes of the ellipses are expressed as

$$(x - x_c) \cos \phi_d + (y - y_c) \sin \phi_d = 0 \tag{2}$$
$$(x - x_c) \sin \phi_d - (y - y_c) \cos \phi_d = 0 \tag{3}$$

The patterns pertaining to horizontal, vertical, and circular polarization are shown in Figure 11.

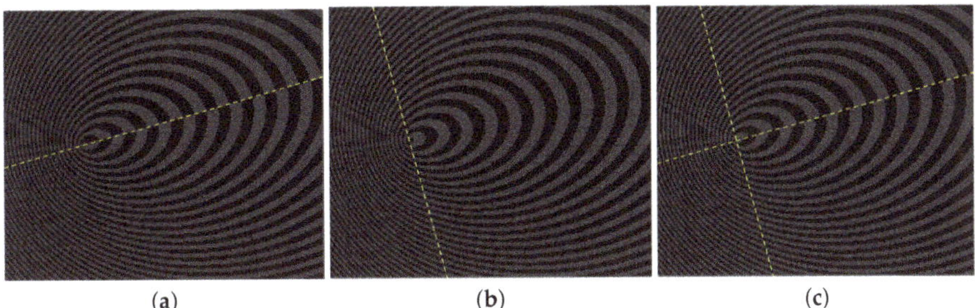

(a)   (b)   (c)

Figure 11. The holographic pattern of the surface-reactance profile needed to form a pencil beam with (**a**) horizontal, (**b**) vertical, and (**c**) circular polarizations.

Polarization reconfigurability can be achieved by feeding the metasurface with two sources instead of one [6]. The surface-reactance modulation function can be obtained as a superposition of the surface-reactance profile needed for each polarization that includes a 90° phase shift between them. Vertically- and horizontally-polarized beams can be formed when the metasurface is excited by the individual sources, and a circularly polarized beam is formed when excited by both sources, as illustrated in Figure 12.

This way, a single metasurface is capable of forming a pencil beam in the desired direction with a horizontal, vertical, or circular polarization. The E-plane radiation patterns demonstrating the polarization reconfigurability is shown in Figure 13.

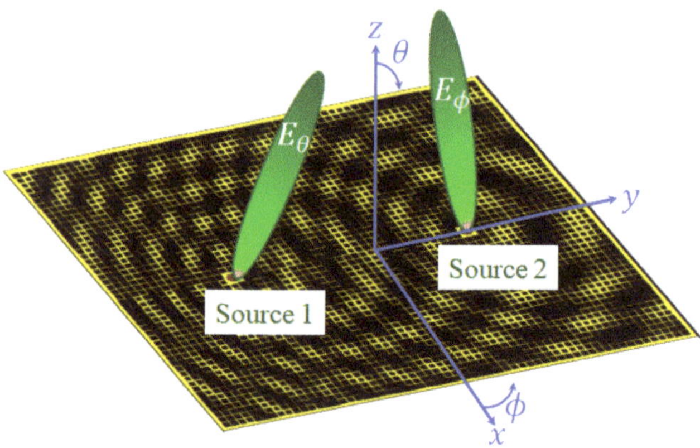

Figure 12. A holographic metasurface, fed by two sources, that forms a pencil beam with three reconfigurable polarization states: vertical, horizontal, and circular polarizations [23].

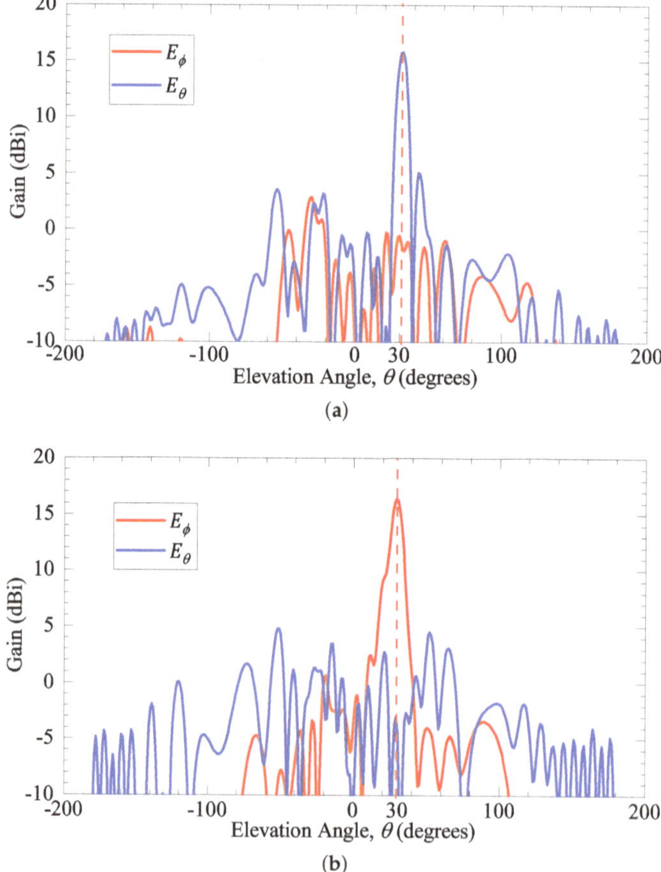

Figure 13. *Cont.*

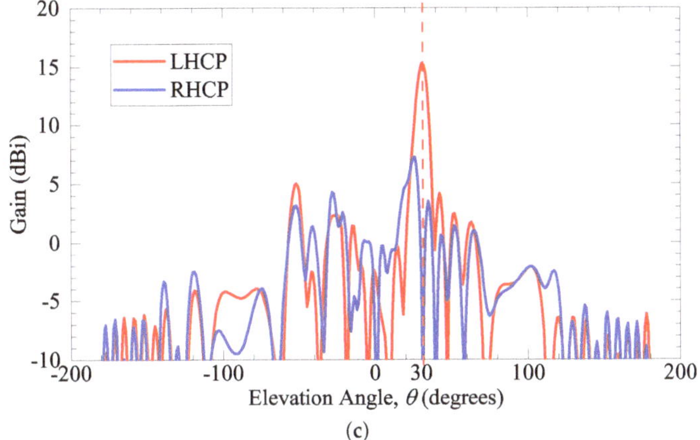

Figure 13. The E-plane radiation patterns obtained at 12 GHz when the antenna is excited by the (**a**) first, (**b**) second, and (**c**) both sources. 12 GHz. The metasurface is realized using Rogers RT/Duroid 5880 substrate of thickness 3.175 mm and a dielectric constant of 2.2. The average surface reactance and the modulation index are 255 and 31 ohms, respectively [23].

The metasurface forms a pencil beam in the desired direction with a vertical, horizontal, and circular polarization when excited by the first, second, and both sources, respectively.

The sense of rotation of the circular polarization depends on the 90° phase lead or lag between the two linear polarizations. A circularly-polarized beam with the other sense of rotation can be formed by externally including a 180° phase shift between the two feeds or by including another source and modulating the metasurface with the superimposition of the surface reactance pertaining to the three sources [6].

Employing metasurfaces and the principle of holography has resulted in a simple configuration of a multi-polarization high-gain LWA, which would otherwise be difficult to achieve. While the presence of multiple sources, yields the metasurface polarization diverse, internal reconfiguration of polarization can be potentially achieved by including switches between the patches. The difference between the surface-reactance profiles that lead to horizontal and vertical polarizations, is the inclusion of a 180° phase difference on either side of the major or minor axis of the ellipses. Electronic switches like PIN diodes can be used to include the 180° phase shift, thus resulting in internal reconfiguration; this yields to a metasurface, which is polarization reconfigurable. While having a switch in every unit cell would be expensive and cumbersome to process, the number of switches can be optimized by a judicious design.

4. A Van Atta Retrodirective Microstrip Antenna Array with Low Backscattering

While the previous two sections demonstrate the application of metasurfaces for surface-wave antenna arrays and leaky-wave antennas, this section focuses on the reconfiguration and improvement of van Atta arrays using AMC-based metasurfaces. In a van Atta reflectarray, the fields reradiated by the antennas are necessary while the fields scattered from the structure (i.e., structural mode) are undesired. Hence, metasurfaces are employed to eliminate these scattered fields from the structural mode of the van Atta retrodirective reflectarray.

4.1. Theory of Passive Retrodirective Arrays

A schematic diagram of a 1-D van Atta array is illustrated in Figure 14, where four array elements are connected using the two transmission lines (i.e., elements '1' and '4' are connected with each-other while element '2' is connected to element '3'). In order to direct the reradiated waves back to the direction of the incident wave, these transmission lines must either be of equal lengths or the difference between their lengths should be a multiple of the wavelength. Since these two transmission lines transfer the signals received by elements '1' and '2' to elements '3' and '4', respectively; the lengths of the transmission lines connecting the antenna element should be kept equal to make the functioning frequency independent. Therefore both transmission lines have equal lengths, $l_1 = l_2 = l$. When a plane wave is incident obliquely upon the array at an angle θ_i, a phase delay Δ between the antenna elements is introduced [44]

$$\Delta = k_0 d \sin \theta_t \tag{4}$$

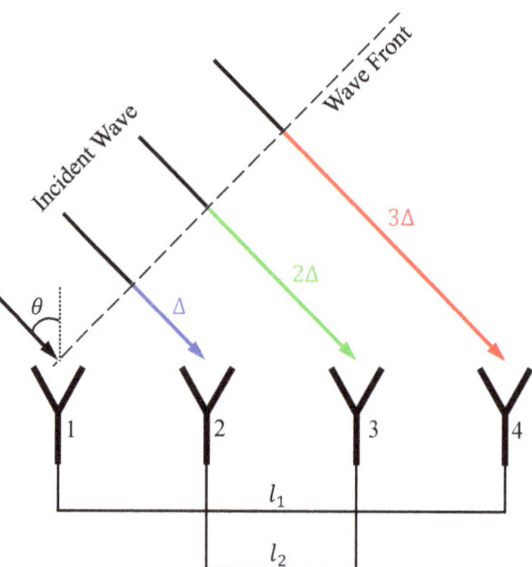

Figure 14. A schematic diagram of a 1-D van Atta retrodirective array where four elements are used in the array.

The monostatic RCS of the retrodirective reflector, based on the reradiated fields only, can be derived following the approach in [38,39] and represented by

$$\sigma^R = \frac{\lambda_0^2}{4\pi} G_0^2 \tag{5}$$

where G_0 is the gain of the array in the principle plane, θ is the incident/elevation angle, and λ_0 is the free-space wavelength. The monostatic RCS (σ^R) of Equation (5) is only based on the fields reradiated ($\mathbf{E^R}$) by the reflector array. However, the total RCS by the reflector array (σ^T) corresponds to the total radiated fields ($\mathbf{E^T}$) which can be determined by adding the reradiated fields ($\mathbf{E^R}$) and the scattered fields by the structure ($\mathbf{E^S}$). Figure 15 shows a comparison of the ideal monostatic RCS between a retrodirective reflector (σ^R) of Equation (5) and a flat perfect electric conductor (PEC) plate, where both have the same size.

The retrodirective response is obtained by having a planar array of 4×4 radiating elements, each of gain $[G_0(\theta)]$ and with $0.5\lambda_0$ spacing between them. Thus, the total size of both the reflector array and the flat PEC plate is $2\lambda_0 \times 2\lambda_0$. It is clear from Figure 15 that the fully passive reflector array can offer a maximum reradiation towards the direction of incidence. However, due to the limitation of the patch antenna towards the grazing angles, the maximum reradiation pattern follows the radiation pattern of the single antenna.

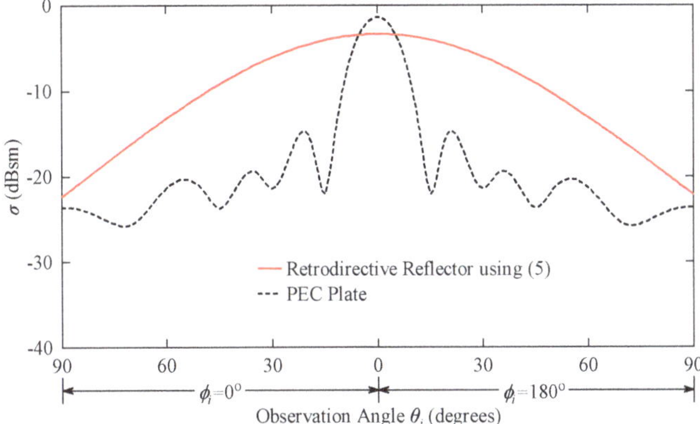

Figure 15. Comparison of the ideal monostatic RCS between a retrodirective reflector (σ^R) obtained by using (5) and a flat PEC.

4.2. Design of the Patch Antenna and Its Feeding Network

In principle, a retrodirective array can be implemented with any kind of antenna element where the bandwidth performance and beamwidth of the array will be directly impacted by the performance parameters of the single radiating element. Due to their design simplicity and their low profile, patch antennas are chosen to design the utilized retrodirective array.

4.2.1. Aperture-Coupled Patch Antenna with Microstrip Line Feeding Network

Figure 16 illustrates the retrodirective reflector that is constructed using a 4×4 finite array of rectangular microstrip-patch antennas.

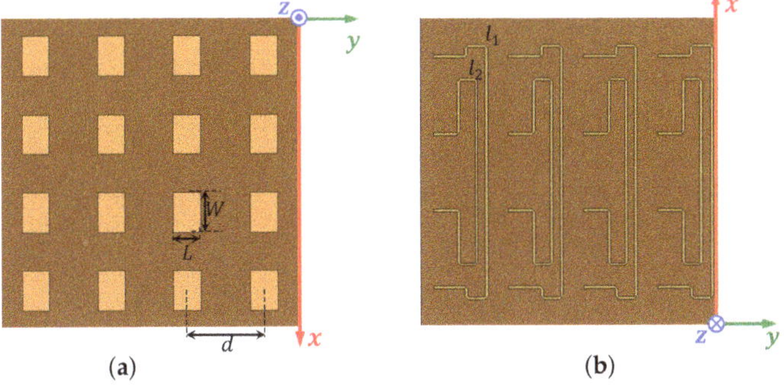

Figure 16. The geometry of the four by four finite array of rectangular microstrip patch antennas that are connected in the van Atta configuration in the xz-plane, (**a**) top. (**b**) bottom.

The elements are interconnected as per the van Atta configuration (i.e., illustrated in Figure 14) in the xz-plane. The feeding network is constructed using a 50 ohms microstrip transmission lines of equal length (i.e., $l = l_1 = l_2$). The dimensions of the square array along the x- and y-direction are 120 mm, corresponding to $2\lambda_0$; λ_0 is the free-space wavelength at 5 GHz. The top side of the structure consists of 16 patch elements resonating at 5 GHz. With reference to Figure 16, the array spacing is $d = 30$ mm ($0.5\lambda_0$), and the patch dimensions are $L = 11$ mm and $W = 15.4$ mm. All the elements are designed on Rogers-RO3006 dielectric substrate (thickness = 1.28 mm, $\varepsilon_r = 6.15$ and $\tan \delta = 0.0022$ at 10 GHz) backed by a PEC.

The above-mentioned design was (modeled in HFSS) illuminated by a plane wave incident along the principal plane (*xz*-plane) defined by $\phi_i = 0°, 180°$ and for $0° \leq \theta_i \leq 90°$. The configuration was simulated with two different lengths (*l*) of the transmission lines that are connecting the antenna elements. The total monostatic RCSs (σ^T) of the retrodirective array for the two transmission lines are depicted in Figure 17. The difference between the two responses is primarily due to the change in the phase of $\mathbf{E^R}$ with respect to $\mathbf{E^S}$ when the length of the transmission lines is changed.

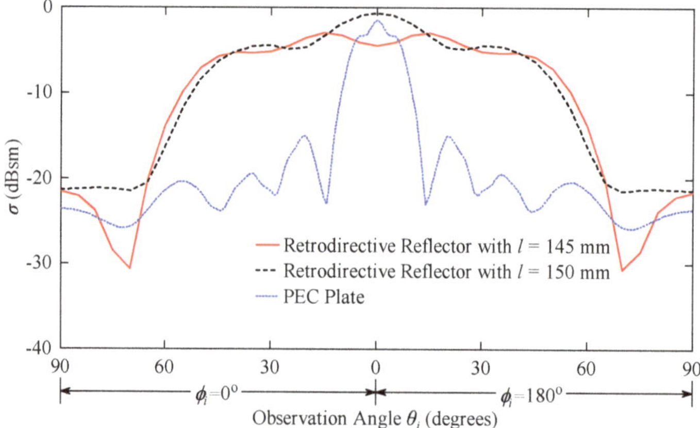

Figure 17. The monostatic RCSs of a retrodirective array with two different transmission line lengths compared to the monostatic RCS of a PEC plate of the same size.

Table 1 summarizes the values of the broadside RCS ($\theta_i = 0°$) of the retrodirective array for the two different lengths of transmission line. It is assumed that the phase of the scattered fields by the array structure ($\mathbf{E^S}$) does not vary significantly with the change of length *l* of the transmission line (this will be verified in the next section). Thus, the total scattered ($\mathbf{E^T}$) will depend on the varying phase introduced by the transmission line. As seen in Table 1, for the first length of $l = 145$ mm, the two components ($\mathbf{E^S}$ and $\mathbf{E^R}$) of $\mathbf{E^T}$ added destructively near broadside incidence, while for the second length of $l = 150$ mm, the two components are in-phase and thus the total RCS is higher than the reradiated σ^R obtained using Equation (5). With the current retrodirective setup (aperture-coupled patch antenna with microstrip line feeding network), it is extremely difficult to extract two components of the total scattered fields. In the next section, it will be shown that the virtual feeding network can be used to identify the two separate components of the total scattered field, and it can permit the targeting and cancellation of one of the components, especially the scattering by the structural mode, which is undesirable.

Table 1. The broadside ($\theta_i = 0°$) RCS of the retrodirective array for two different coaxial cable lengths.

	$l = 145$ mm	$l = 150$ mm	Analytical Using (5)
σ (dBsm)	−4.42	−0.63	−3.24

4.2.2. Aperture-Coupled Patch Antenna with Virtual Feeding Network

An alternative method for simulating the retrodirective reflector explained in Section 4.2.1 is to use a virtual feeding network. The basic concept of a virtual feeding network is to capture how much voltage (magnitude and phase) is received by each antenna element, and then feed the measured voltages to the respective antenna elements following the van Atta configuration shown in Figure 14. All the radiating elements of the array are terminated with an impedance equal to the input impedance of the antenna (ideally 50 ohms). Then, at the termination point, the received voltage from the illuminating plane wave is determined by integrating along the electric field lines around the microstrip lines, similar to quasi-static simulators. This integration is performed based on the mode of propagation, which is Quasi-Transverse Electromagnetic (Quasi-TEM) as illustrated in Figure 18 [45]. The above-mentioned process is achieved by using a co-simulation between the full-wave simulator (Ansys Electromagnetic Suite/HFSS), and a Python code that is written to integrate the electric fields and then feed those calculated voltages again to their respective elements in the full-wave simulator. Figure 19 shows the structure that was configured in HFSS. Here, each of the antenna ports (i.e., transmitting antennas) is terminated with an impedance that is equal to the input impedance of the corresponding connected antenna (i.e., receiving antenna) to replicate the connection between them.

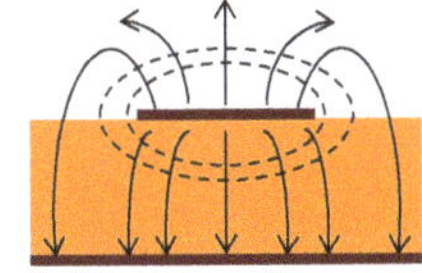

— E-fields   - - - H-fields

Figure 18. Electric and magnetic field lines for the TEM mode of a coaxial line.

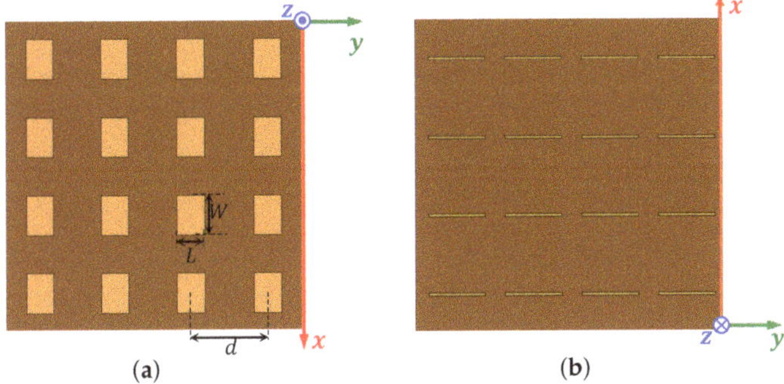

(a)   (b)

Figure 19. (**a**) Top and (**b**) bottom geometries of the 4 × 4 finite array of rectangular microstrip-patch antennas that are terminated and virtually connected in the van Atta configuration in the *xz*-plane.

Similar to the arrangement of Section 4.2.1, the structure is illuminated by a plane wave incident in the principal plane (*xz*-plane) defined by $\phi_i = 0°, 180°$ and for $0 \leq \theta_i \leq 90°$. Again, by using the Python script, the voltage captured by each antenna element at each incident angle is recorded and fed again to their counterpart reradiating element. This process allows for the decoupling of the total scattered fields $\mathbf{E^T}$ into the two main components: reradiated fields $\mathbf{E^R}$ and structural mode scattering $\mathbf{E^S}$; their corresponding RCSs are shown in Figure 20. It can be shown that the phase of $\mathbf{E^S}$ is primarily dependent on the structure, and it is not impacted by the phase introduced by the connecting cables. However, the phase of $\mathbf{E^R}$ can be altered based on the connecting cables. Thus, to illustrate the impact of the phase difference (Δ_{RS}) at the broadside (θ_i) between $\mathbf{E^S}$ and $\mathbf{E^R}$, the total RCS by the retrodirective array is shown in Figure 20 when $\Delta_{RS} = 0°$ and $180°$, respectively.

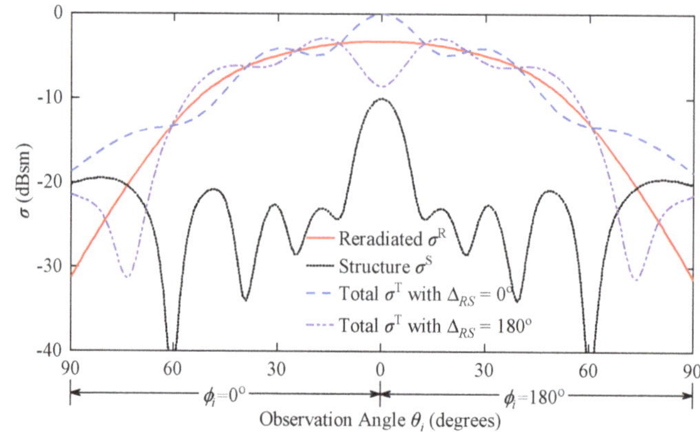

Figure 20. The monostatic RCSs based on the reradiated, scattered, and the total fields when the phase difference (Δ_{RS}) between $\mathbf{E^R}$ and $\mathbf{E^S}$ is $0°$ and $180°$.

4.3. Retrodirective Reflector with Low Backscattering Using Metamaterials

It was shown in the previous section that the scattered fields from the structure introduce ripples in the total RCS by the reflector, especially near the broadside, where the RCS of the structure is large. Ideally, $\mathbf{E^S}$ can be canceled by a judiciously synthesized metasurface, like that of the widely-known concept of checkerboard metasurfaces [46].

Design of Metasurfaces Based on AMCs for RCS Reduction of Retrodirective Array

The RCS of the array structure can be reduced by placing an AMC, whose reflection is similar in magnitude but with a $180°$ phase difference compared to that of $\mathbf{E^S}$. As shown in Figure 20, the RCS of the structural mode is around -10 dBsm and the phase of its reflected fields is around $130°$. Thus, an AMC surface of square patches is designed where their width and unit cell dimensions are selected to have a reflection phase of around $-50°$ and an RCS of -10 dBsm, similar to the monostatic RCS corresponding to the structural mode of the antenna.

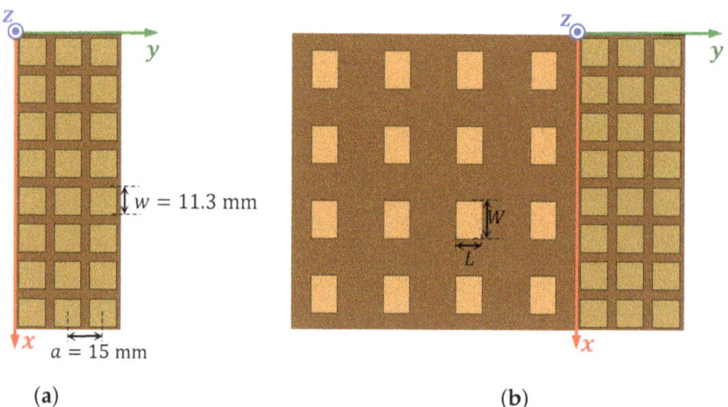

Figure 21. Geometries of (**a**) the finite array of the AMC and (**b**) the 4×4 array of rectangular microstrip patch antennas combined with the finite AMC array for canceling the RCS of the structure.

The geometries of the synthesized finite sized AMC metasurface are depicted in Figure 21a. As a result, as illustrated in Figure 21b, the final design of the proposed retrodirective reflector is obtained by placing the original antenna array of Figure 19 adjacent to the synthesized AMC of

Figure 21a. This configuration will introduce fields by the AMC with a phase difference of 180° compared to the phase of the scattered fields by the antenna structure.

This design is then illuminated by following a similar arrangement to the retrodirective reflector without the AMC, as explained previously in Section 4.2. The monostatic RCSs due to the antenna structure, reradiated fields, and the total fields are shown in Figure 22. It can be seen that regardless of the phase introduced by the connecting cables that impact the phase of the reradiated fields only, the monostatic RCS caused by the structure σ^S is reduced at and near broadside. This causes the total RCS to have smaller ripples and the performance becomes less sensitive to the interference by the scattered fields of the structure.

The RCS, which corresponds to the fields scattered by the antenna structure, can be reduced by placing a metasurface with a similar RCS and 180° phase difference with respect to the reflected fields of the antenna array. The range of the phase difference can be relaxed to achieve an RCS reduction of at least 10 dB when the phase difference is maintained within $(180 \pm 37)°$ [46,47]. Using this criterion, the RCS of various antennas has been reduced for more than 10 dB over a wide frequency band [46,48,49]; furthermore, a 10-dB RCS reduction is not necessary here. It is sufficient that the RCS of the scattered fields is lower than that of the reradiated fields by a certain factor based on the acceptable level of the ripples in the total field. Consequently, the phase-difference criterion of $(180 \pm 37)°$ can be relaxed further, and a much broader operational bandwidth can be attained. In this model, the metasurface is designed to achieve the RCS-reduction at the operating frequency of the van Atta reflector to validate the concept, which can be applied over a broader frequency range.

The scanning capability of the van Atta reflector depends on the overall gain of the array (G_0) as given by Equation (5), where G_0 is determined by the radiation pattern of the single element and the spacing between them. Thus, individual radiating elements with broad beamwidths result in a reflector that responds to larger incident angles. However, antennas with broad beamwidths usually have a lower gain. As a figure of merit, the 10-dB beamwidth of the array is considered as its scanning range. In this design, the monostatic RCS plotted in Figure 22 shows that the 10-dB beamwidth of the designed array is about 120° in the principle scanning plane (i.e., xz-plane).

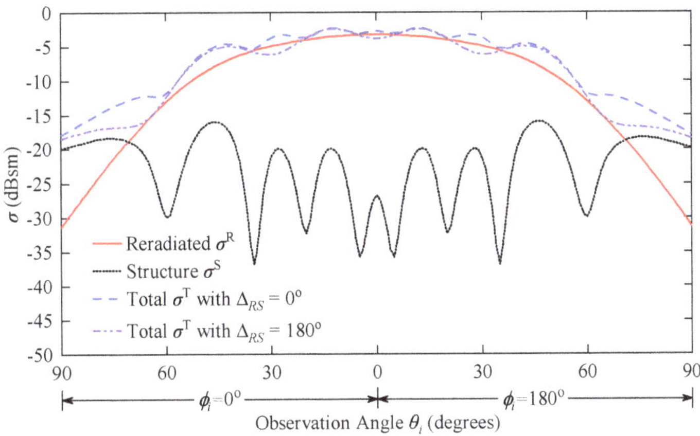

Figure 22. The monostatic RCSs based on the reradiated, scattered, and the total fields when the phase difference (Δ_{RS}) between $\mathbf{E}^R$ and $\mathbf{E}^S$ is 0° and 180°.

5. Conclusions

The application of metasurfaces in the reconfiguration and enhancement of the performance of surface-wave antenna arrays, holographic leaky-wave antennas, and retrodirective arrays is illustrated. Square ring-element arrays embedded within metasurfaces are proposed for linear and circular polarizations. The coupling between the ports and their effect on the broadside radiation are discussed and highlighted. For practicality, parallel feed networks are incorporated with the two proposed designs. With the incorporation of metasurfaces, a high realized gain and a high aperture efficiency, compared to a conventional square ring array, are observed. A prototype was fabricated and a good agreement between the simulations and measurements was demonstrated.

Metasurface LWAs designed using the holographic principle are reviewed. In particular, polarization-diverse holographic metasurfaces that can form a pencil beam with a horizontal, vertical, or circular polarization are discussed. A potential technique to incorporate electronic reconfigurability is proposed.

As the third area of illustration, metasurfaces are employed to improve the monostatic RCS performance of retrodirective arrays. A co-simulation method that allows for decoupling the total scattered fields is introduced. Such a simulation allows the identification and cancellation of the undesired component in the scattered fields (those scattered by the antenna structure). It is shown that this work enables the use of retrodirective arrays comprised of simple patch antennas while simultaneously addressing the issue of backscattering by using metasurfaces based on AMCs.

Author Contributions: Conceptualization, S.R. and C.A.B.; methodology, M.A., M.A.A. and S.R.; software, M.A., M.A.A., and S.R.; validation, C.R.B., C.A.B. and A.Y.M.; formal analysis, S.R., A.Y.M. and C.A.B.; investigation, M.A., M.A.A. and S.R.; resources, M.A., M.A.A. and S.R.; data curation, M.A., M.A.A., S.R. and C.R.B.; writing—original draft preparation, M.A., M.A.A. and S.R.; writing—review and editing, C.A.B., S.R. and A.Y.M.; visualization, M.A., M.A.A. and S.R. All authors have read and agreed to the published version of the manuscript.

Funding: This research received no external funding.

Conflicts of Interest: The authors declare no conflict of interest.

References

1. Bukhari, S.S.; Vardaxoglou, J.; Whittow, W. A Metasurfaces Review: Definitions and Applications. *Appl. Sci.* **2019**, *9*, 2727. [CrossRef]
2. Pfeiffer, C.; Grbic, A. Metamaterial Huygens surfaces: Tailoring wave fronts with reflectionless sheets. *Phys. Rev. Lett* **2013**, *110*, 197401. [CrossRef] [PubMed]
3. Maci, S.; Minatti, G.; Casaletti, M.; Bosiljevac, M. Metasurfing: Addressing Waves on Impenetrable Metasurfaces. *IEEE Antennas Wirel. Propag. Lett.* **2011**, *10*, 1499–1502. [CrossRef]
4. Fong, B.H.; Colburn, J.S.; Ottusch, J.J.; Visher, J.L.; Sievenpiper, D.F. Scalar and tensor holographic artificial impedance surfaces. *IEEE Trans. Antennas Propag.* **2010**, *58*, 3212–3221. [CrossRef]
5. Ramalingam, S.; Balanis, C.A.; Birtcher, C.R.; Pandi, S.; Shaman, H.N. Axially Modulated Cylindrical Metasurface Leaky-Wave Antennas. *IEEE Antennas Wirel. Propag. Lett.* **2018**, *17*, 130–133. [CrossRef]
6. Ramalingam, S.; Balanis, C.A.; Birtcher, C.R.; Pandi, S.; Shaman, H.N. Polarization-diverse holographic metasurfaces. *IEEE Antennas Wirel. Propag. Lett.* **2019**, *18*, 264–268. [CrossRef]
7. Modi, A.Y.; Balanis, C.A.; Birtcher, C.R.; Shaman, H.N. Novel Design of Ultrabroadband Radar Cross Section Reduction Surfaces Using Artificial Magnetic Conductors. *IEEE Trans. Antennas Propag.* **2017**, *65*, 5406–5417. [CrossRef]
8. Modi, A.Y.; Alyahya, M.A.; Balanis, C.A.; Birtcher, C.R. Metasurface-Based Method for Broadband RCS Reduction of Dihedral Corner Reflectors With Multiple Bounces. *IEEE Trans. Antennas Propag.* **2020**, *68*, 1436–1447. [CrossRef]
9. Ramalingam, S.; Balanis, C.A.; Birtcher, C.R.; Pandi, S. Analysis and Design of Checkerboard Leaky-Wave Antennas With Low Radar Cross Section. *IEEE Open J. Antennas Propag.* **2020**, *1*, 26–40. [CrossRef]
10. Alharbi, M.S.; Balanis, C.A.; Birtcher, C.R. Hybrid Circular Ground Planes for High-Realized-Gain Low-Profile Loop Antennas. *IEEE Antennas Wirel. Propag. Lett.* **2018**, *17*, 1426–1429. [CrossRef]
11. Alharbi, M.S.; Balanis, C.A.; Birtcher, C.R. Performance Enhancement of Square-Ring Antennas Exploiting Surface-Wave Metasurfaces. *IEEE Antennas Wirel. Propag. Lett.* **2019**, *18*, 1991–1995. [CrossRef]
12. Costa, F.; Luukkonen, O.; Simovski, C.R.; Monorchio, A.; Tretyakov, S.A.; de Maagt, P.M. TE Surface Wave Resonances on High-Impedance Surface Based Antennas: Analysis and Modeling. *IEEE Trans. Antennas Propag.* **2011**, *59*, 3588–3596. [CrossRef]

13. Yang, F.; Aminian, A.; Rahmat-Samii, Y. A Novel Surface-Wave Antenna Design Using a Thin Periodically Loaded Ground Plane. *Microw. Opt. Technol. Lett.* **2005**, *47*, 240–245. [CrossRef]
14. Yang, F.; Rahmat-Samii, Y.; Kishk, A. Low-Profile Patch-Fed Surface Wave Antenna With a Monopole-Like Radiation Pattern. *IET Microw. Antennas Propag.* **2007**, *1*, 261–266. [CrossRef]
15. Feng, G.; Chen, L.; Xue, X.; Shi, X. Broadband Surface-Wave Antenna With a Novel Nonuniform Tapered Metasurface. *IEEE Antennas Wirel. Propag. Lett.* **2017**, *16*, 2902–2905. [CrossRef]
16. Lin, F.H.; Chen, Z.N. Low-Profile Wideband Metasurface Antennas Using Characteristic Mode Analysis. *IEEE Trans. Antennas Propag.* **2017**, *65*, 1706–1713. [CrossRef]
17. Yang, W.; Chen, S.; Che, W.; Xue, Q.; Meng, Q. Compact High-Gain Metasurface Antenna Arrays Based on Higher-Mode SIW Cavities. *IEEE Trans. Antennas Propag.* **2018**, *66*, 4918–4923. [CrossRef]
18. Zhu, H.L.; Cheung, S.W.; Chung, K.L.; Yuk, T.I. Linear-to-Circular Polarization Conversion Using Metasurface. *IEEE Trans. Antennas Propag.* **2013**, *61*, 4615–4623. [CrossRef]
19. Ta, S.X.; Park, I. Low-Profile Broadband Circularly Polarized Patch Antenna Using Metasurface. *IEEE Trans. Antennas Propag.* **2015**, *63*, 5929–5934. [CrossRef]
20. Wu, Z.; Li, L.; Li, Y.; Chen, X. Metasurface Superstrate Antenna With Wideband Circular Polarization for Satellite Communication Application. *IEEE Antennas Wirel. Propag. Lett.* **2016**, *15*, 374–377. [CrossRef]
21. Ta, S.X.; Park, I. Compact Wideband Circularly Polarized Patch Antenna Array Using Metasurface. *IEEE Antennas Wirel. Propag. Lett.* **2017**, *16*, 1932–1936. [CrossRef]
22. Patel, A.M.; Grbic, A. A printed leaky-wave antenna based on a sinusoidally-modulated reactance surface. *IEEE Trans. Antennas Propag.* **2011**, *59*, 2087–2096. [CrossRef]
23. Ramalingam, S. Impedance Modulated Metasurface Antennas. Ph.D. Dissertation, Arizona State University, Tempe, AZ, USA, 2020.
24. Minatti, G.; Caminita, F.; Casaletti, M.; Maci, S. Spiral leaky-wave antennas based on modulated surface impedance. *IEEE Trans. Antennas Propag.* **2011**, *59*, 4436–4444. [CrossRef]
25. Pandi, S.; Balanis, C.A.; Birtcher, C.R. Design of Scalar Impedance Holographic Metasurfaces for Antenna Beam Formation with Desired Polarization. *IEEE Trans. Antennas Propag.* **2015**, *63*, 3016–3024. [CrossRef]
26. Minatti, G.; Caminita, F.; Martini, E.; Sabbadini, M.; Maci, S. Synthesis of modulated-metasurface antennas with amplitude, phase, and polarization control. *IEEE Trans. Antennas Propag.* **2016**, *64*, 3907–3919. [CrossRef]
27. Bodehou, M.; Martini, E.; Maci, S.; Huynen, I.; Craeye, C. Multibeam and Beam Scanning With Modulated Metasurfaces. *IEEE Trans. Antennas Propag.* **2020**, *68*, 1273–1281. [CrossRef]
28. Li, M.; Tang, M.; Xiao, S. Design of a LP, RHCP and LHCP Polarization-Reconfigurable Holographic Antenna. *IEEE Access* **2019**, *7*, 82776–82784. [CrossRef]
29. Van Atta, L.C. Electromagnetic Reflector. U.S. Patent 2,908,002, 6 October 1959.
30. Sharp, E.; Diab, M. Van atta reflector array. *IEEE Trans. Antennas Propag.* **1960**, *8*, 436–438. [CrossRef]
31. Appel-Hansen, J. A van atta reflector consisting of half-wave dipoles. *IEEE Trans. Antennas Propag.* **1996**, *14*, 694–700. [CrossRef]
32. Chung, S.J.; Chang. K. A retrodirective microstrip antenna array. *IEEE Trans. Antennas Propag.* **1998**, *46*, 1802–1809. [CrossRef]
33. Li, Y.; Jandhyala, V. Design of retrodirective antenna arrays for short-range wireless power transmission. *IEEE Trans. Antennas Propag.* **2012**, *60*, 206–211. [CrossRef]
34. Wang, X.; Sha, S.; He, J.; Guo, L.; Lu, M. Wireless power delivery to low-power mobile devices based on retro-reflective beamforming. *IEEE Antennas Wirel. Propag. Lett.* **2014**, *13*, 919–922. [CrossRef]
35. Ettorre, M.; Alomar, W.A.; Grbic, A. Radiative wireless power-transfer system using wideband, wide-angle slot arrays. *IEEE Trans. Antennas Propag.* **2017**, *65*, 2975–2982. [CrossRef]
36. Tseng, W.J.; Chung, S.B.; Chang, K. A planar van atta array reflector with retrodirectivity in both e-plane and h-plane. *IEEE Trans. Antennas Propag.* **2000**, *48*, 173–175. [CrossRef]
37. Hester, J.G.D.; Tentzeris, M.M. Inkjet-printed flexible mm-wave van-atta reflectarrays: A solution for ultralong-range dense multitag and multisensing chipless rfid implementations for iot smart skins. *IEEE Trans. Microw. Theory Tech.* **2016**, *64*, 4763–4773. [CrossRef]
38. Christie, S.; Cahill, R.; Buchanan, N.B.; Fusco, V.F.; Mitchell, N.; Munro, Y.V.; Maxwell-Cox, G. Rotman lens-based retrodirective array. *IEEE Trans. Antennas Propag.* **2012**, *60*, 1343–1351. [CrossRef]
39. Ettorre, M.; Alomar, W.A.; Grbic, A. 2-d van atta array of wideband, wideangle slots for radiative wireless power transfer systems *IEEE Trans. Antennas Propag.* **2018**, *66*, 4577–4585. [CrossRef]

40. Sievenpiper, D.F. High-Impedance Electromagnetic Surfaces. Ph.D. Dissertation, University of California, Los Angeles, CA, USA, 1999.
41. Yang, F.; Rahmat-Samii, Y. Microstrip antennas integrated with electromagnetic band-gap (EBG) structures: A Low mutual coupling design for array applications. *IEEE Trans. Antennas Propag.* **2003**, *51*, 2936–2946. [CrossRef]
42. Quarfoth, R. Anisotropic Artificial Impedance Surfaces. Ph.D. Dissertation, University of California, San Diego, CA, USA, 2014.
43. Balanis, C.A. *Advanced Engineering Electromagnetics*, 2nd ed.; Wiley: Hoboken, NJ, USA, 2012.
44. Balanis, C.A. *Antenna Theory: Analysis and Design*, 4th ed.; John Wiley & Sons: Hoboken, NJ, USA, 2016.
45. Pozar, D.M. *Microwave Engineering*, 4th ed.; John Wiley & Sons: Hoboken, NJ, USA, 2011.
46. Modi, A.Y.; Balanis, C.A.; Birtcher, C.R.; Shaman, H.N. New Class of RCS-Reduction Metasurfaces Based on Scattering Cancellation Using Array Theory. *IEEE Trans. Antennas Propag.* **2019**, *67*, 298–308. [CrossRef]
47. Chen, W.; Balanis, C.A.; Birtcher, C.R. Checkerboard EBG surfaces for wideband radar cross section reduction. *IEEE Trans. Antennas Propag.* **2015**, *63*, 2636–2645. [CrossRef]
48. Liu, Y.; Li, K.; Jia, Y.; Hao, Y.; Gong, S.; Guo, Y.J. Wideband RCS Reduction of a Slot Array Antenna Using Polarization Conversion Metasurfaces. *IEEE Trans. Antennas Propag.* **2016**, *64*, 326–331. [CrossRef]
49. Zheng, Y.; Gao, J.; Zhou, Y.; Ziangyu, C.; Yang, H.; Li, S.; Li, T. Wideband Gain Enhancement and RCS Reduction of Fabry–Perot Resonator Antenna With Chessboard Arranged Metamaterial Superstrate. *IEEE Trans. Antennas Propag.* **2018**, *67*, 590–599. [CrossRef]

electronics

Article

C-Band and X-Band Switchable Frequency-Selective Surface

Umer Farooq [1], Adnan Iftikhar [1], Muhammad Farhan Shafique [2], Muhammad Saeed Khan [1], Adnan Fida [1], Muhammad Juanid Mughal [1] and Dimitris E. Anagnostou [3,*]

[1] Department Electrical and Computer Engineering, COMSATS University Islamabad, Islamabad 45550, Pakistan; dr.umerfarooq094@gmail.com (U.F.); adnaniftikhar@comsats.edu.pk (A.I.); mskj786@hotmail.com (M.S.K.); adnan_fida@comsats.edu.pk (A.F.); junaid.mughal@comsats.edu.pk (M.J.M.)

[2] Center for Advanced Studies in Telecommunication (CAST), COMSATS University Islamabad, Islamabad 45550, Pakistan; farhan.shafique@comsats.edu.pk

[3] Department of Electrical Engineering, Heriot-Watt University, Edinburgh EH14 4AS, UK

* Correspondence: d.anagnostou@hw.ac.uk

Abstract: This paper presents a highly compact frequency-selective surface (FSS) that has the potential to switch between the X-band (8 GHz–12 GHz) and C-band (4 GHz–8 GHz) for RF shielding applications. The proposed FSS is composed of a square conducting loop with inward-extended arms loaded with curved extensions. The symmetric geometry allows the RF shield to perform equally for transverse electric (TE), transverse magnetic (TM), and 45° polarizations. The unit cell has a dimension of 0.176 λ_0 and has excellent angular stability up to 60°. The resonance mechanism was investigated using equivalent circuit models of the shield. The design of the unit element allowed incorporation of PIN diodes between adjacent elements for switching to a lower C-band spectrum at 6.6 GHz. The biasing network is on the bottom layer of the substrate to avoid effects on the shielding performance. A PIN diode configuration for the switching operation was also proposed. In simulations, the PIN diode model was incorporated to observe the switchable operation. Two prototypes were fabricated, and the switchable operation was demonstrated by etching copper strips on one fabricated prototype between adjacent unit cells (in lieu of PIN diodes) as a proof of the design prototypes. Comparisons among the results confirmed that the design offers high angular stability and excellent performance in both bands.

Keywords: active FSS; C-band; frequency-selective surface (FSS); spatial filters; switchable; X-band

Citation: Farooq, U.; Iftikhar, A.; Shafique, M.F.; Khan, M.S.; Fida, A.; Mughal, M.J.; Anagnostou, D.E. C-Band and X-Band Switchable Frequency-Selective Surface. *Electronics* **2021**, *10*, 476. https://doi.org/10.3390/electronics10040476

Academic Editor: Bang Chul Jung

Received: 1 January 2021
Accepted: 13 February 2021
Published: 17 February 2021

Publisher's Note: MDPI stays neutral with regard to jurisdictional claims in published maps and institutional affiliations.

1. Introduction

Nowadays, urban environments are saturated with electromagnetic energy produced by the wireless devices used in our daily routines. This energy is meant for limited numbers of people, but everyone is exposed to it. Apart from causing interferences for nearby devices, it may also has adverse long-term effects on human health, causing tissue damage as well as the development of tumors in the worst cases [1–5]. In secured installations, such as intensive care units (ICUs), security and military installations, operation theaters, and cardiac care units (CCUs), communication can be restricted to frequency spectra below the C-band (4 GHz–8 GHz), in which only 3G and 4G mobile connectivity is maintained. This limitation very much suffices for basic communication needs. In these systems, all higher-frequency signals, such as those in the upper 802.11ac WLAN band, as well as in some ISM bands, are desired to be rejected.

In addition, the X-band (8–12 GHz) (used in space communications and radars) is another high-frequency and high-power electromagnetic (EM) spectrum that is present in the environment and can endanger human beings. The mitigation of RF radiation in restricted zones where security is a prime concern is often required by security agencies. Therefore, to mitigate unwanted EM radiation and electromagnetic interference (EMI) in controlled environments, several methods have been practiced, such as shielding an environment/device using metal foils or metallic coating [6]. These conventional shielding methods block useful signals along with unwanted signals, and also add additional cost, weight, and bulkiness to devices [6,7].

On the contrary, frequency-selective surfaces (FSSs) [8], due to their single- and multi-band frequency suppression characteristics, have been extensively used in military installations, secure locations such as jails, and various other sites [9,10]. In the literature, several configurations of FSSs have been reported to mitigate EMI for GSM, WLAN/Wi-Fi, and X-band signals in a single- or multi-mode operation [11–13]. In addition, frequency-reconfigurable and tunable FSSs that are capable of shielding more than two frequency bands have also been reported [14–18]. Various methods, such as mechanical deformation, material tuning, circuit biasing, and micro-fluidic techniques, have been employed to alter the frequency response of FSSs [19]. However, mechanical deformations are difficult to implement due to complexities related to their physical controls. Similarly, material tuning causes surfaces to suffer from environmental conditions, such as humidity and changes in temperature. Circuit biasing is a frequently employed method for achieving reconfiguration. Varactor and PIN diodes have been used [7,11,12]. The active shields realized using the above-mentioned techniques are extremely useful in extracting the optimal frequency spectra as per the requirements of end users. Nonetheless, the major limitations associated with these active FSSs are their polarization dependence and poor shielding performance at different angles of incidence [14–16].

The authors of [16] proposed a Jerusalem-cross-based tunable FSS with a unit cell size of 27.5 mm to shield 900 and 1800 MHz, whereas varactors were used to tune lower bands. However, the authors did not report the FSS's behavior for transverse electric (TE), transverse magnetic (TM) polarizations, or the shielding stability at normal and oblique incidence angles. In addition, a tunable meta-material FSS with 10.5 mm size was tuned from 3 to 3.5 GHz band by using varactors on both sides of the substrates [17], but polarization independence of the FSS was not established to demonstrate its suitable candidacy for practical applications. An FSS geometry with a unit cell size of 10.5 mm operating at 5.8 GHz was made reconfigurable by using a mechanical setup that altered the gaps between the FSS elements in order to attain frequency-reconfigurable operation [18]. Nevertheless, the performance of the FSS proposed in [18] at oblique incidence angles was not reported. In addition, reconfigurable FSSs using two techniques—one was based on using diodes to change the current distribution, and the second was based on mechanical rotation of the FSS—were proposed in [20]. The realizations of both techniques were performed only in simulations, which, therefore, may raise performance concerns in the practical realization of the proposed FSS shields. In [21], a venetian-blind-based shutter mechanism was used to achieve reconfigurability in an FSS, but the shutter mechanism was employed manually. Similarly, a reconfigurable dual-band-stop FSS using a liquid crystal polymer was proposed in [22] without in-depth analysis or discussions on the practical realization or angular stability. In the existing literature, only one flexible and reconfigurable FSS with angular stability has been proposed so far, which can be found in [23]. The FSS proposed in [23] was designed on flexible plastic and fabricated using an additive manufacturing process, whereas varactor diodes with a bias line on the same layer of the FSS were used to achieve frequency reconfigurability. The reported FSS offered better shielding performance by virtue of the costly additive manufacturing process, and it required extra effort to place the FSS on the host dielectric, which caused additional design bulkiness. In addition, the effect of strong coupling between RF and DC currents because of the placement of the FSS and DC bias lines on the same layers resulted in a poor shielding performance by the FSS.

It can be seen in the above literature review and discussion that, presently, FSS geometries with reconfigurability have sizes of about $2.7\lambda_0$ or more for X-band shielding [12], which is mainly because the multi-band, tunable, and reconfigurable shielding operations increase the size of the structure, and it is very hard to sustain the polarization independence and angular stability, as reported in the literature [14–18]. In this very context, the objective of this work is to propose a highly compact and efficient FSS shield for X-band applications with an option of switching to C-band suppression. The novelty is justified because such compact FSS designs for switchable X- and C-band shielding are not available in the literature. In this paper, it is successfully demonstrated that the proposed FSS design has a stable frequency response for both the TE/TM and 45° polarizations at normal and oblique angles of incidence of up to 60°. The remainder of the paper is organized as follows: The FSS unit cell modeling and simulation results are presented in Section 2. Mathematical and equivalent circuit models of the design are presented in Section 3, whereas a parametric analysis of the design is discussed in Section 4. The proposed mechanism of active switching to the C-band is discussed in Section 5. A comparison of the simulated and measured results for various incident angles is shown in Section 6, followed by the conclusion in the last section.

2. Proposed Passive FSS Unit Cell

A schematic view of the proposed square-loop FSS with inward-extended quarter-circular arms and with optimized dimensions is shown in Figure 1a. The proposed design was realized on a single layer of a low-cost FR4 laminate with a thickness h = 1.6 mm, relative permittivity ϵ_r = 4.4, and loss tangent tanδ = 0.02. An FSS unit cell was first modeled and optimized in a 3D full-wave EM simulator (Ansys HFSS) [24]. Initially, a conventional square loop was simulated, and its frequency response was observed. Afterwards, to miniaturize the proposed design and increase the angular stability of the proposed FSS, inward-extended quarter-circular arms were incorporated to achieve the maximum shielding effectiveness (SE) in the desired band. Furthermore, an inward-extended and curved arm was introduced to tune the shielding frequency by keeping the overall unit cell size at 4.8 mm × 4.8 mm. For the idealization of the FSS unit cell as an EM shield with infinite periodicity in the x- and y-axes, a Floquet port analysis [25] with master and slave boundaries (linked boundaries) was performed to achieve the desired shielding effectiveness for the X-band. An illustration of the Floquet analysis along with the port declaration and periodic boundary setup is presented in Figure 1b–d. Fundamentally, the Floquet modes impinge on an EM plane wave normal to the FSS, i.e., the propagation direction is set by the frequency, phasing, and geometry of the periodic structure. The simulator solution included a model decomposition with information on the performance of the radiating structure. The impinging EM waves in the Floquet modes then interrelated with the S-matrix, and the S-parameter results were computed for an infinite array under periodic boundary conditions (shown in Figure 1b–d). Next, the dB value of $|S_{21}|$ was plotted against the frequency to observe the SE of the designed FSS. The SE is often defined as the ratio of forward power transmitted by the source (P_f) to power received at the termination (P_r). Mathematically,

$$SE_{dB} = 10log_{10}\frac{P_f}{P_r},\qquad(1)$$

which can be further simplified and interpreted in terms of $|S_{21}|$ as:

$$SE_{dB} = -20log_{10}\,|\,S_{21}\,|\,.\qquad(2)$$

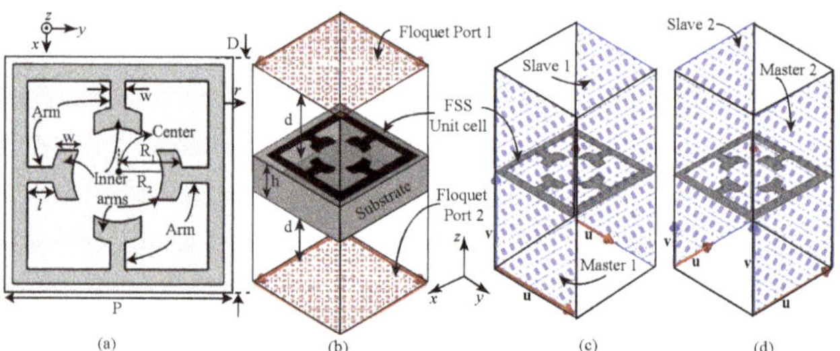

Figure 1. Unit cell geometry of the frequency-selective surface (FSS) and simulation setup for the unit cell investigation. (**a**) Geometry of the proposed miniaturized and symmetric FSS unit cell with optimized dimensions: P = 4.8 mm, L = 4.2 mm, w = 0.3 mm, R_1 = 0.8 mm, and R_2 = 0.8 mm; (**b–d**) FSS unit cell model with periodic boundary conditions and Floquet port assignments.

3. Equivalent Circuit Model (ECM)

To accurately predict the resonance behavior and SE of the proposed FSS, an equivalent circuit model (ECM) was extracted using the lumped element model theory [26–31], and the values were determined using the coupled-line theory [26,27]. A plane wave from port 1 (shown in Figure 1b) propagates through free space and illuminates the shield. This illumination results in surface currents at the resonant frequency around the closed loop of the FSS [27]. The free-space impedance is modeled as Z_0 (377 Ω), whereas the square-loop conducting structure in the proposed FSS unit cell is modeled using inductances. For the symmetrical modeling, inductance is divided into half on both sides of the inner arms, which divide each leg of the square loop into two parts, as shown in Figure 2a. The conductive nature of the inward-extended arms is also modeled as an inductor L_{arm}, whereas the coupling capacitance C_c is incorporated to attribute the coupling between the inner extended arms. The construction and mapping of each conducting strip of the proposed FSS on the equivalent lumped elements are depicted in Figure 2a. In addition, C_{fss} in the lumped model is included to represent the coupling between the adjacent cells in a periodic structure of the proposed FSS. Since the distance between the inner extended arms is greater than the distance between the adjacent cells, the combined coupling capacitance C_c can be neglected during the subsequent circuit simplification (Figure 2c). An illustration of the lumped element model extracted from Figure 2a is depicted in Figure 2b. The impedance of the dielectric used in the unit cell is modeled by Z_T; however, it may be neglected because of the incomparable thickness of the substrate [27].

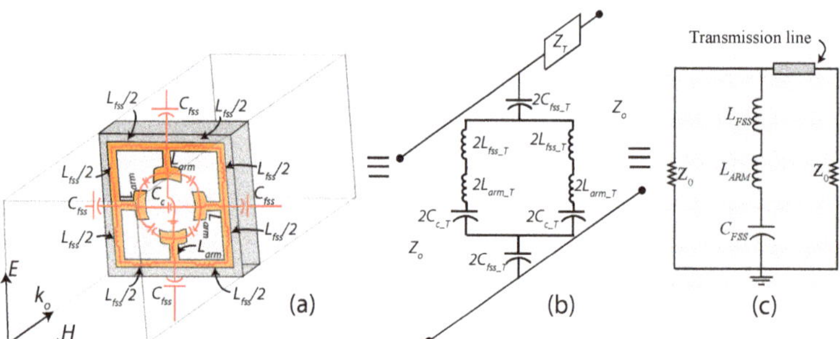

Figure 2. (**a**) Mapping and construction of the equivalent lumped element model of the proposed FSS unit cell, (**b**) the equivalent circuit model (ECM), and (**c**) the simplified ECM of the proposed FSS unit cell.

It can be seen in Figure 2c that the total impedance of the proposed FSS (Z_{FSS}) is a series and parallel combination of the inductances and capacitances shown in Figure 2b. Therefore, after simplification, the final construction of the ECM of the proposed FSS is effectively a series LC circuit in parallel with the free-space impedance at both ends. Mathematically, the FSS impedance Z_{FSS} is expressed as [26,27]:

$$Z_{FSS} = X_{L_{FSS}} + X_{L_{ARM}} + X_{C_{FSS}}, \tag{3}$$

where $X_{L_{FSS}}$ is the reactance corresponding to the complete length of the square loop. $X_{C_{FSS}}$ represents the coupling between the two adjacent unit cells, and $X_{L_{ARM}}$ represents the lengths of the inner extended arms, as shown in Figure 2c. By replacing the equations for the reactances in Equation (3), we get:

$$Z_{FSS} = j\omega L_{FSS} + j\omega L_{ARM} + \frac{1}{j\omega C_{FSS}}. \tag{4}$$

Rearranging Equation (4),we obtain

$$Z_{Fss} = \frac{j^2\omega^2 L_{FSS}C_{FSS} + j^2\omega^2 L_{ARM}C_{FSS} + 1}{j\omega C_{FSS}}. \tag{5}$$

Further simplification of Equation (5) results in Equation (6):

$$Z_{FSS} = \frac{1 - (\omega^2 C_{FSS})[L_{FSS} + L_{ARM}]}{j\omega C_{FSS}}. \tag{6}$$

The shielding frequency f_r of the ECM in Equation (6) is obtained by equating the numerator to zero [29]. This implies that

$$f_r = \left(\frac{1}{2\pi}\right)\left(\sqrt{\frac{1}{C_{FSS}(L_{FSS} + L_{ARM})}}\right). \tag{7}$$

It can be interpreted from Equations (6) and (7) that the combined capacitive and inductive reactances corresponding to the distance between adjacent unit cells and conductive strips of the FSS, respectively, cancel each other at the shielding frequency f_r. The cancellation of the inductive and capacitive reactances results in a short circuit in the ECM and, consequently, reflects all EM energy that strikes the surface of the shield. In addition, for $f < f_r$, the capacitive nature of the lumped LC circuit dominates the resonance mechanism. Conversely, when the impedance of the series LC circuit is inductive at $f > f_r$, a higher shielding frequency f_r with dominant inductive impedance—as compared to the capacitive impedance of the lumped LC circuit—results. Equation (7) also shows that various combinations can be formulated using different values of inductances and capacitances through the lumped circuit theory [27]. Therefore, for the shielding frequency of 11 GHz, when optimized through 3D EM simulations, the values of the equivalent lumped elements are finalized at C_{FSS} = 0.065 pF, L_{FSS} = 2.4 nH, and L_{ARM} =0.8 nH. For the verification of the ECM, values of the reactances obtained from Equation (7) were used to simulate Figure 2c in a circuit simulator [32]. A comparison between the SE of the unit cell modeled in the 3D EM software (Figure 1) and the ECM is presented in Figure 3. It shows that the EM simulation results are in good agreement with the circuit simulation results.

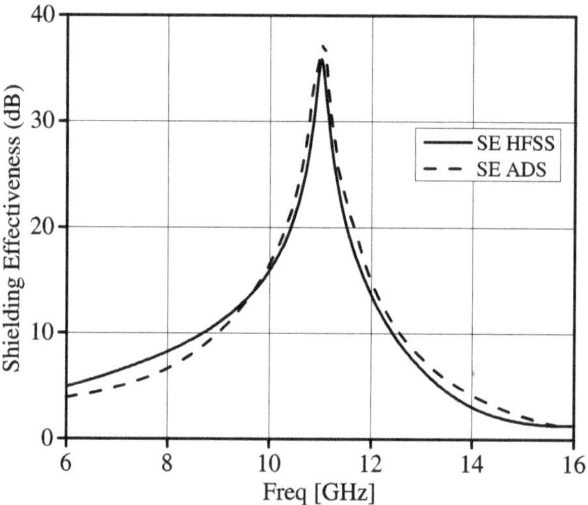

Figure 3. A comparison of the shielding effectiveness (SE) results of the EM simulation and circuit simulation.

4. Parametric Analysis

After the simulation and circuit modeling of the unit cell, a detailed parametric analysis was performed to observe the behavior of the proposed FSS with variations in the geometric parameters and dielectric constant of the substrate. Specifically, variations in the shielding frequency were investigated by varying (1) unit element size, (2) the extended arm length l, (3) the trace width w of the unit element (shown in Figure 1a) , and (4) the dielectric constant ϵ_r of the substrate. It was observed that changing the FSS parameters altered the overall inductance L_{FSS} and capacitance C_{FSS} of the FSS structure. Mathematically, inductance is expressed as [31]:

$$L_{FSS} = \mu_0 \left(\frac{1}{2\pi} \right) log_{10} \left(\frac{1}{sin \frac{\pi w}{2D}} \right), \tag{8}$$

where μ_0 is the permeability of free space, l is the length of inner extended arms of the FSS unit cell, D is the overall length of the square loop, and w is the width of the FSS conductor. Equation (8) shows that inductance L_{FSS} has direct dependence on D, whereas an increase in width w decreases the inductance of the FSS structure. The direct and indirect dependence of the length and width of the FSS was verified by performing a parametric analysis on the length and width of the FSS unit cell in the EM simulator. Figure 4a shows the effect on shielding frequency with respect to the change in FSS length. It can be seen in Figure 4a that changing the arm length while keeping other parameters constant shifts the shielding frequency of the proposed FSS. Similarly, performing a parametric analysis on the width w of the FSS unit cell showed that inductance decreases as w increases; thus, shielding frequency shifts to the higher side, as shown in Figure 4b. Furthermore, the effect on the shielding frequency f_r was observed by changing inter-element spacing r. It was observed that an increase in r decreases the capacitance of the FSS units and results in a higher shielding frequency, as is evident in Figure 4c and the relation given in Equation (9).

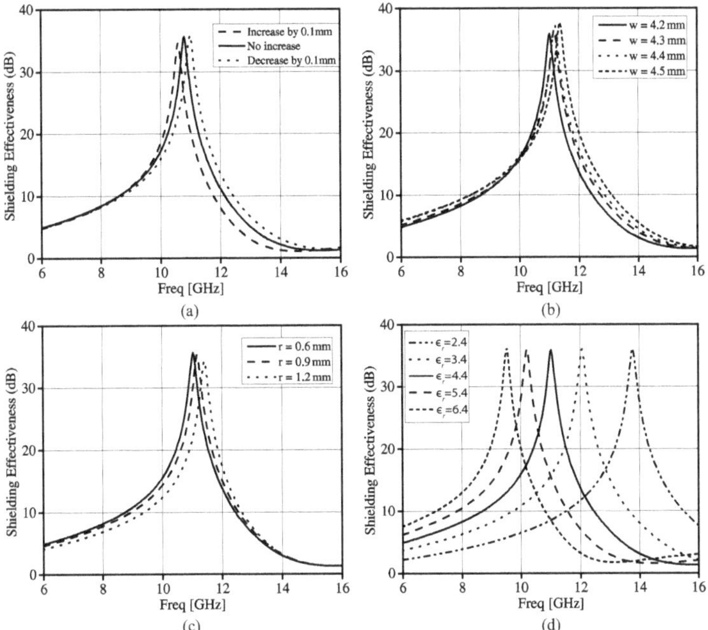

Figure 4. Parametric analysis of geometric parameters and their effects on f_r by changing the (**a**) arms' length l, (**b**) strip width w, (**c**) element spacing r, and (**d**) dielectric constant ϵ_r.

$$C_{FSS} = \epsilon_0 \epsilon_{eff} \left(\frac{2D}{\pi} \right) log_{10} \left(\frac{1}{sin \frac{\pi r}{2D}} \right). \tag{9}$$

The dielectric constant ϵ_r was also varied to analyze the FSS's performance, and the corresponding results are shown in Figure 4d. It was observed that, as the value of ϵ_r starts increasing, the f_r tends to decrease. This is because the square-loop length D is inversely related to ϵ_{eff}; mathematically, it is represented as [12]:

$$D \approx \frac{c}{4 f_r \sqrt{\epsilon_{eff}}}, \tag{10}$$

where ϵ_{eff} is the effective permittivity and is approximated as $\epsilon_{eff} = \dfrac{(\epsilon_r + 1)}{2}$.

Angular Stability of the FSS at Normal and Oblique Angles of Incidence

The shielding performance of the proposed FSS under a normal angle of incidence for TE and TM polarizations is presented in Figure 5. Overall, identical and stable shielding responses at the X-band were observed for both the TE and TM polarizations because of the symmetrical structure of the FSS.

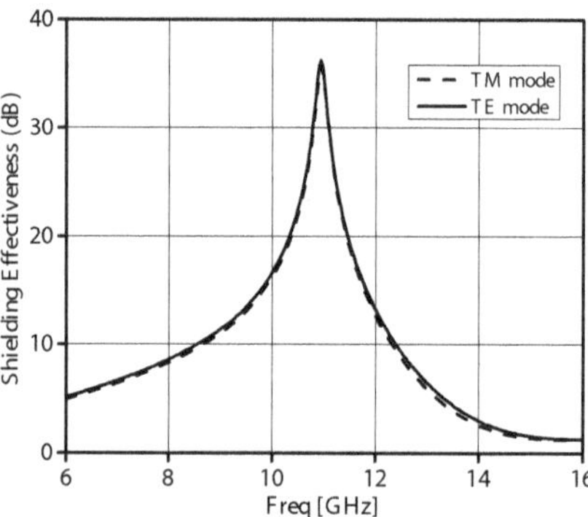

Figure 5. SE of the FSS for the transverse magnetic (TM) and transverse electric (TE) polarizations at a normal angle of incidence.

Moreover, the performance of the FSS at various angles of incidence for both the TE and TM polarizations was also analyzed to demonstrate the practicability of the FSS. For performance analysis at various angles, two Floquet ports with TE and TM modes at oblique angles (0°, 30°, and 60°) were used in the simulation model of the FSS. The results showed that the proposed FSS had a stable frequency response for TM polarization at angles varying from 0° to 60°, which correspond to higher modes. In the conventional square-loop FSS unit cell, the absence of the conductive patterns in the inner area resulted in poor angular stability. However, in the proposed FSS unit cell, four inner conductive arms with curved strips covered the inner area and provided greater angular stability for the EM waves arriving from the different angles. Thus, the extended inner arms of the proposed square-loop FSS not only tuned the shielding frequency (f_r), but also provided stability to the FSS at various angles of incidence. From Figure 6a , it can be seen that the SE increased from 35 to 39 dB as the angle of incidence was increased from 0° to 60° for TM polarization, with less than 1% change in f_r. However, for the TE polarization, slight shifts of 1.2% and 1% in f_r were observed as the angle of incidence was increased from 0° to 60°, respectively. In addition, for the TE polarization, f_r showed a stable response, as depicted in Figure 6b.

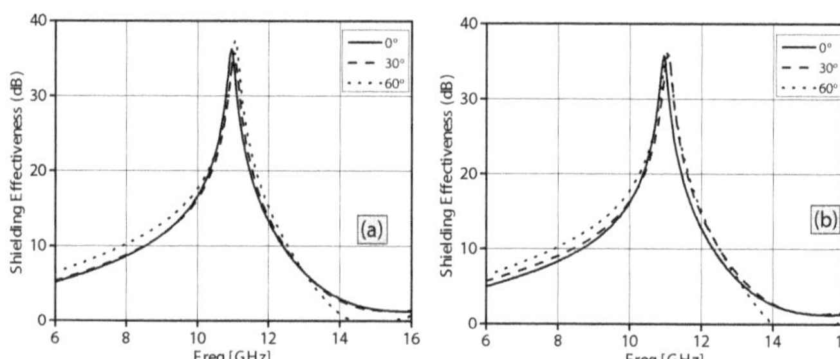

Figure 6. Shielding response of the proposed FSS at oblique angles (0°, 30°, and 60°): (**a**) TM polarization and (**b**) TE polarization.

In addition, a stable frequency response of the proposed FSS at oblique angles and changes in bandwidth at oblique angles for the TM and TE polarizations were observed because of the variation in the surface impedance at oblique angles. A 15 dB bandwidth comparison of both polarizations is shown in Table 1. It was found that bandwidth increases with an increase in the angle of incidence.

Table 1. Bandwidth comparison.

Incidence Angle	TM Polarization	TE Polarization
0°	1.90 GHz	2.05 GHz
15°	1.94 GHz	2.16 GHz
30°	2.11 GHz	2.18 GHz
45°	2.35 GHz	2.68 GHz
60°	3.27 GHz	3.25 GHz

5. Proposed Active FSS

The proposed FSS unit cell shown in Figure 1 can be reconfigured using PIN diodes (SMP-1322-079LF [33])to suppress the C-band (6.6 GHz) instead of the X-band. Reconfigurability was achieved by incorporating PIN diodes in the gaps between the adjacent unit cells of the FSS. To demonstrate this reconfigurability, a finite 3×3 element FSS was simulated, as shown in Figure 7a. The PIN diodes were modeled as equivalent RLC boundaries between adjacent cells. The "OFF" and "ON" states of the PIN diodes in the ECM are also shown in Figure 7a. The PIN diodes act as an open circuit in the "OFF" state. However, for the "ON" state, an additional inductance appears between the unit cells, resulting in the reduction of the shielding resonant frequency to the C-band. The value of C_{fss} (Figure 2a) also changes, since there is a capacitance between the adjacent elements, but due to the imperfect nature of metallic strips, a low value of capacitance C_{fss} still exists, which results in the reduction of the resonant frequency. The response of the reconfigurable FSS is shown in Figure 7b. It can be confirmed from the figure that the FSS operates in either the X-band or C-band with the PIN diodes in the "OFF" state or "ON" state, respectively. The shielding resonant frequency for the C-band is centered at 6.6 GHz with a 15 dB bandwidth of 2.75 GHz.

Furthermore, to elaborate on the reconfigurability operation of the proposed FSS, induced surface current distributions for a 3×3 element shield are plotted in Figure 8 for both bands. The colors of the arrows show the intensity of the surface currents, while current flow is represented by the direction of arrows. It can be observed in Figure 8a that the "OFF" state of PIN diodes isolated the flow of current between adjacent elements and resulted in shielding of the X-band. However, for the "ON" state of diodes, the current flowed through them and increased the effective length, resulting in suppression of the C-band. For the TE mode, which is shown in Figure 8b, the PIN diodes in the "ON" state carried a strong current on the left and right sides, whereas a low current intensity can be observed on the top and bottom diodes of the unit element. The PIN diodes were modeled as equivalent lumped elements and simulated using RLC boundary conditions. However, in practical realization, a biasing circuitry is necessary in order to operate these PIN diodes. In this regard, a biasing mechanism was proposed to operate these diodes under forward or reverse bias conditions. The biasing network was designed on the back side of the shield with thin conductive lines, as shown in Figure 9. These conductive lines were connected to the diodes on the top layer with conductive vias; two diodes were connected to one conductive line, and and the other two were connected to the other conductive line in each unit element. These pairs were isolated using capacitors to avoid short circuiting. Since the biasing circuit was on the bottom side of the FSS, the effect of the DC current on the RF current, which was on top layer, would be at its minimum, and the shielding effectiveness of the FSS would remain the same as that in the simulation results. In addition, the capacitors incorporated on each arm of the FSS in the top layer in the proposed biasing circuitry would avoid short circuiting because of any DC current leakage and would pass the RF current. However, a negligible effect on shielding effectiveness may exist because of the stray capacitance. Therefore, it can be argued that the incorporated capacitors will have a negligible effect on the overall shielding effectiveness of the proposed FSS.

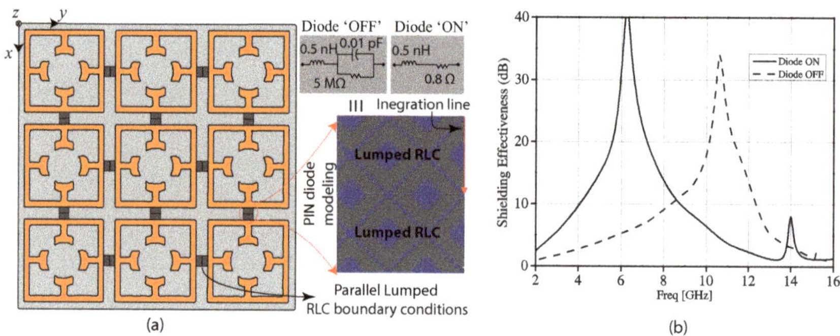

Figure 7. (**a**) Illustration of the reconfigurable variant of the proposed FSS and (**b**) the simulated SE of the proposed FSS unit cell with PIN diodes in the "OFF" and "ON" states.

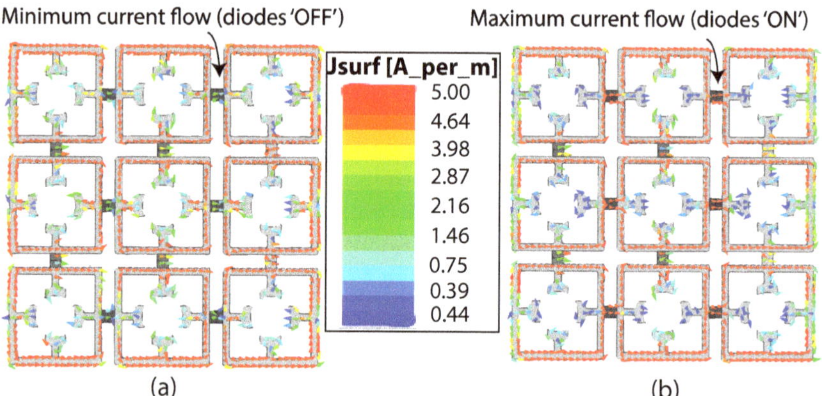

Figure 8. Surface current distribution of the reconfigurable FSS in the TE mode with (**a**) a PIN diode in the "OFF" state at 11 GHz and (**b**) a PIN diode in the "ON" state at 6.6 GHz.

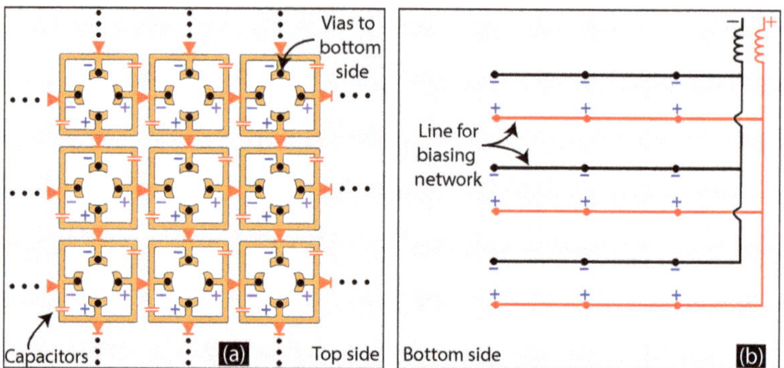

Figure 9. Biasing network topology for PIN diode switching; (**a**) PIN diodes' locations and vias for connections and (**b**) the biasing network on the back side with RF chokes.

6. Measured Results

To validate the proposed designs, two separate shields were fabricated. One shield had no connections between unit elements, representing the "OFF" state of the PIN diodes, and the other one had shorting strips, representing the "ON" state. Both shields were fabricated on an FR4 substrate with $\epsilon_r = 4.4$, $\tan\delta = 0.02$, and 1.6 mm thickness. Each shield had 54 × 52 elements. Pictures of the FSS prototypes fabricated with and without metallic strips to mimic PIN diodes' "ON" and "OFF" states, respectively, are shown in Figure 10a,b. The overall size of the fabricated prototype was $9.78\lambda_0 \times 9.32\lambda_0$ at a frequency of 11 GHz. To measure the shielding effectiveness of the fabricated prototype, a rectangular aperture with dimensions 240 mm × 240 mm was created in an aluminum sheet with size 260 mm × 250 mm (greater than $6\lambda_0$). This was done to ensure that the EM plane wave was only transmitted through the FSS fixed in the aperture. Two identical broadband horn antennas (model#LB-20265) [34] were used as a transmitter and receiver. A block diagram of the experimental setup used for the measurements is shown in Figure 10c. Furthermore, to ensure the efficacy and accuracy, a two-stage measurement process was adopted where, in the first stage, the reference measurement of $|S_{21}|$ was carried out through the aperture without a shield. This reference measurement was performed to measure the free-space transmission through the aperture. Then, In the second stage, the shields were placed in the aperture and measured with reference to the calibration. Similarly, for angular stability measurements, the setup was rotated at different angles.

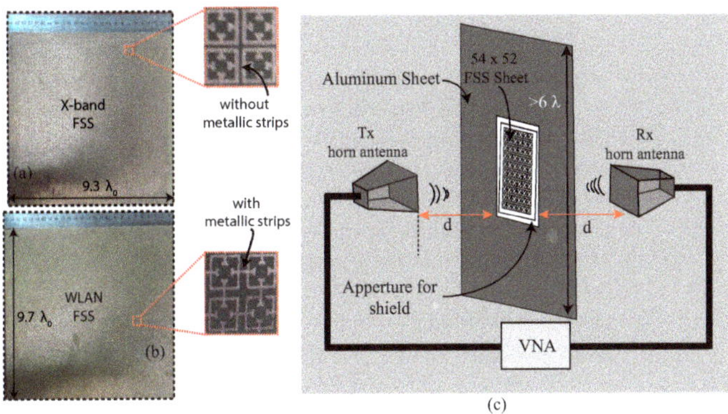

Figure 10. FSSs fabricated with 54 × 52 unit cells on a 260 mm × 250 mm FR4 board: (**a**) FSS without metallic strips for the X-band shielding and (**b**) FSS with metallic strips for C-band shielding. (**c**) Illustration of the experimental setup for all measurements.

Performance at Normal and Oblique Angles of Incidence

A comparison of the simulated and measured results at angles of incidence of 0°, 30°, and 60° for the TE polarization is shown in Figure 11A. The results for the TM polarization at different incident angles should be the same as those in Figure 11 because of the symmetry in the design geometry. It can be seen in Figure 11A that the finite FSS sheet without metallic strips (in PIN diodes' "OFF" state) had a stable angular response for the TE/TM polarizations at the X-band. The comparison showed that the measured results are in good agreement with those of the simulations. The fabricated FSS offered 30 dB SE for the X-band. The angular stability for the 45° polarization was also measured and compared with the simulated results, as shown in Figure 11B. The results show that the proposed FSS has a stable shielding response for oblique angles of incidence.

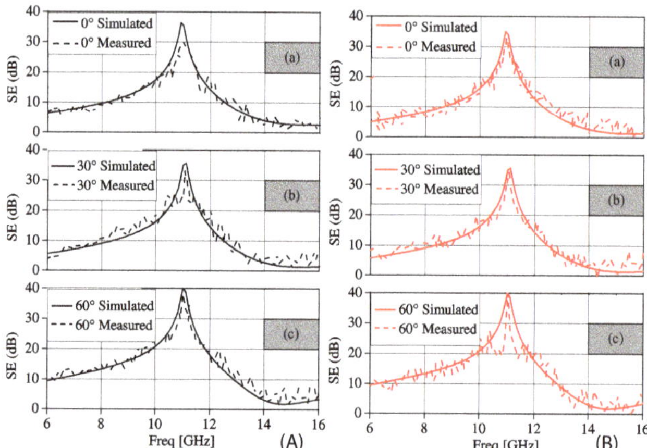

Figure 11. (**A**) Angular stability analysis of the proposed FSS for the TE/TM polarizations at (**a**) 0°, (**b**) 30°, and (**c**) 60°. (**B**) Angular stability analysis for the 45° polarization at (**a**) 0°, (**b**) 30°, and (**c**) 60°.

Furthermore, the performance of the proposed FSS with metallic strips (PIN diodes' "ON" state) in suppressing the C-band was also measured for TE/TM and 45° polarization with normal and oblique angles of incidence (0°, 30°, and 60°). The proposed FSS with metallic strips offered 40 dB shielding for the TE/TM polarizations at oblique angles, as is evident in Figure 12A,B. Overall, the stable shielding response and good agreement between the simulation and measured results confirm the employability of the proposed highly compact FSS in applications such as radomes, hospitals, and security areas. It can also be concluded from Figures 11 and 12 that a good agreement between the simulation and measured results exists when the PIN diode equivalent circuit model is used in simulations and metallic strips are used in the measurement to mimic the PIN diodes' "ON" state. The purpose of the PIN diodes in the "ON" state is to provide a continuous current path between the adjacent unit cells of the FSS, whereas the "OFF" state of the PIN diodes disconnects current flow between the unit cells of the FSS. Therefore, the use of the metallic strips for the practical demonstration of the FSS serves the same purpose as that of the "ON" state of the PIN diodes.

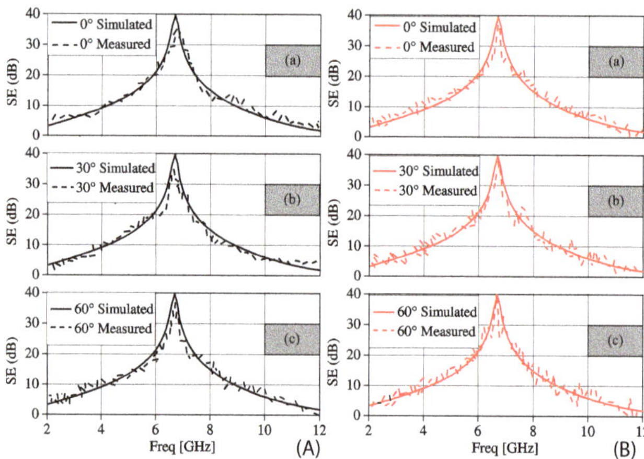

Figure 12. (**A**) Performance comparison of the FSS with metallic strips for the TE/TM polarization at (**a**) 0°, (**b**) 30°, and (**c**) 60°. (**B**) Performance comparison for 45° polarization at (**a**) 0°, (**b**) 30°, and (**c**) 60°.

Finally, a comparison table (Table 2) is also included in the manuscript to highlight the novelty of this work in comparison with the other reported reconfigurable/tunable active FSSs. A comparison is sketched between the existing reconfigurable/tunable active FSSs and the proposed active FSS. In addition, a column of reconfigurable techniques is also included in the comparison table to describe the merits and demerits of the previously reported FSSs. In addition to the use of PIN diodes and varactor diodes, mechanical tuning through manual rotation of the FSS has also been reported in literature. However, these manual rotations are not autonomous and require a large mechanical setup for rotation to achieve reconfigurable operation of an FSS. On the other hand, the PIN diodes and associated biasing circuitry are used on the top-side substrate where the shielded structure is designed, which significantly affects the shielding performance of the FSS. In addition, the tunability of an FSS is demonstrated using varactor diodes. Nonetheless, effect of shielding effectiveness still exists because of the coplanar configuration of the FSS, varactor diode, and associated biasing circuitry. The practical demonstration of the proposed FSS validated the employability of the shield in C- and X-band shielding applications, such as in airborne radars and parabolic reflectors for satellite communications.

Table 2. Performance comparison with previously published literature.

Ref.	Reconfigurable/ Tunable Frequencies	Angular Stability/ Polarization	Technique Used	Merits/Demerits
[18]	3.5 GHz, 5 GHz, and 5.8 GHz	Not reported	Mechanical tuning by varying distance	• Customized mechanical setup • Not easily applicable for other FSSs • Polarization stability analysis is not possible • A motor is required to rotate the shafts
[19]	12.42 GHz	Up to 30°	PIN diodes	• Demonstrated only isolation and no isolation in diodes' ON and OFF states, respectively • Results demonstrated for only 1×3 array
[21]	3.2 GHz, 4.75 GHz, and 5.2 GHz	Up to 60°	Manual rotation of blinds	• The rotation of blinds to control EM waves for FSS may not be practical • The FSS structure is difficult to control for frequency reconfigurability
[23]	3.5 GHz–5.7 GHz	Not reported	Varactor Diodes	• Additive manufacturing technique is costly • Conformability analysis is provided
[35]	5.7 GHz–9.8 GHz	Up to 30°	PIN Diodes	• One parallel bias circuit configuration • Designed on a flexible substrate and offers only 30° angular stability
This work	6.6 GHz and 11 GHz	Up to 60°	PIN Diodes	• Biasing circuit on the bottom layer is proposed • Structure is simple and can be printed using cost-effective printing techniques

7. Conclusions

In this paper, a highly compact FSS with unit cell dimensions of 4.8 mm × 4.8 mm ($0.176\lambda_0 \times 0.176\lambda_0$) was realized in simulations and practically demonstrated using measurements. The optimization of the shielding effectiveness (SE) as a function of the geometric parameters and the substrate's electrical properties was performed in simulations and analytically. It was successfully demonstrated that the proposed FSS provides 25–35 dB SE for the X-band and can be reconfigured to the C-band using PIN diodes installed between the adjacent unit elements. A complete parametric study at normal and oblique angles of incidence for the TE and TM polarizations is also presented to demonstrate the shielding stability of the proposed FSS with and without metallic strips to mimic the "ON" and "OFF" states of the PIN diodes. Furthermore, an equivalent circuit model (ECM) and mathematical relations for interpreting the dependency of the shielding frequency on the FSS geometry were also formulated using the transmission line theory. The FSS was fabricated and measured to highlight its employability in practical applications. The simulated and measured results agreed well and validated the design concept of the compact FSS for X- and C-band shielding applications.

Author Contributions: U.F. performed the complete simulations. A.I. performed fabrications and measurements. M.F.S. and M.J.M. provided the idea. Post-processing of the results and their comparison were done by M.S.K., A.F., and M.F.S., and D.E.A. assisted in overall paper management, idea development, and manuscript writing. All authors contributed equally to this work. All authors have read and agreed to the published version of the manuscript.

Funding: This work was supported in part by the EU H2020 Marie Skłodowska-Curie Individual Fellowship under Grant # 840854 (VisionRF), and by COMSATS University Islamabad under the COMSATS Research Grant Program (No. 16-65/CGRP/CUI/ISB/18/849).

Conflicts of Interest: The authors declare no conflict of interest.

References

1. Chen, N.Z.; Cai, A.; See, T.S.P.; Qing, X.; Chia, M.Y.W. Small planar UWB antennas in proximity of the human head. *IEEE Trans. Microw. Tech.* **2006**, *54*, 1846–1857. [CrossRef]
2. Kara, A.; Bertoni, H.L. Effect of people moving near short-range indoor propagation links at 2.45 GHz. *J. Comm. Net.* **2006**, *8*, 286–289. [CrossRef]
3. Singh, S.; Kapoor, N. Occupational EMF exposure from radar at X and Ku frequency band and plasma catecholamine levels. *Bioelectromagnetic* **2015**, *36*, 444–450. [CrossRef]
4. Hongwei, H. The effect of human activities on 2.4 GHz radio propagation at home environment. In Proceedings of the 2nd IEEE International Conference on Broadband Network Multimedia Technology, Beijing, China, 18–20 October 2009; pp. 95–99.
5. Cleary, S.F.; Garber, F.; Liu, L.M. Effects of X-band microwave exposure on rabbit erythrocytes. *Bioelectromagnetics* **1982**, *3*, 453–466. [CrossRef]
6. Munk B A. *Frequency Selective Surfaces-Theory and Design*; John Wiley & Sons: Hoboken, NY, USA, 2000.
7. Yadav, S.; Jain, C.P.; Sharma, M.M. Smart phone frequency shielding With penta-bandstop FSS for Security and electromagnetic health applications. *IEEE Trans. Electromagn. Compat.* **2018**, *61*, 887–892. [CrossRef]
8. Farooq, U.; Shafique, M.F.; Mughal, M.J. Polarization insensitive dual band frequency selective surface for RF shielding through glass windows. *IEEE Trans. Electromagn. Compat.* **2020**, *62*, 93–100. [CrossRef]
9. Farooq, U.; Iftikhar, A.; Shafique, M.F.; Mughal, M.J. A miniaturized and polarization insensitive FSS and CFSS for dual band WLAN applications. *Elservier AEU-Int. J. Electron. Comm.* **2019**, *105*, 124–134. [CrossRef]
10. Farooq, U. Ultraminiaturised Polarisation Selective Surface (PSS) for dual-band Wi-Fi and WLAN shielding applications. *IET Microwaves Antennas Propag.* **2020**, *14*, 1514–1521. [CrossRef]
11. Yu, Y.; Chiu, C.; Chiou, Y.; Wu, T. A Novel 2.5-Dimensional ultra-miniaturized-element frequency selective surface. *IEEE Trans. Antennas. Propag.* **2014**, *62*, 3657–3663. [CrossRef]
12. Nauman, M.; Saleem, R.; Rashid, A.K.; Shafique, M.F. A miniaturized flexible frequency selective surface for X-band applications. *IEEE Trans. Electromagn. Compat.* **2016**, *58*, 419–428. [CrossRef]
13. Bagci, F.; Mulazimoglu, C.; Can, S.; Karakaya, E.; Yilmaz, A.E.; Akaoglu, B. A glass based dual band frequency selective surface for protecting systems against WLAN signals. *Int. J. Electron. Commun. (AEU)* **2017**, *82*, 426–434. [CrossRef]
14. Ford, K.L.; Roberts, J.; Zhou, S.; Fong, G.; Rigelsford, J. Reconfigurable frequency selective surface for use in secure electromagnetic buildings. *Electron. Lett.* **2013**, *49*, 861–863. [CrossRef]
15. Majidzadeh, M.; Ghobadi, C.; Nourina, J. A reconfigurable frequency-selective surface for dual-mode multi-band filtering applications. *Int. J. Electron.* **2016**, *104*, 369–381. [CrossRef]
16. Masud, M.M.; Ijaz, B.; Ullah, I.; Braaten, B. A Compact Dual-Band EMI Metasurface shield with an actively tunable polarized lower band. *IEEE Trans. Electromagn. Compat.* **2012**, *54*, 1182–1185. [CrossRef]
17. Bayatpur, F.; Sarabandi, K. Tunable metamaterial frequency-selective surface with variable modes of operation. *IEEE Trans. Microw. Theory Tech.* **2009**, *57*, 1433–1438. [CrossRef]
18. Sivasamy, R.; Moorthy, B.; Kanagasabai, M.; Samsingh, V.R.; Alsath, M.G.N. A wideband frequency tunable FSS for electromagnetic shielding applications. *IEEE Trans. Electromagn. Compat.* **2018**, *60*, 280–283. [CrossRef]
19. Lee, I.-G.; Kim, Y.-J.; Park, Y.-B.; Chun, H.-J.; Hong, I.-P. Design of X-band reconfigurable frequency selective surface with high isolation. *IEICE Electron. Express* **2016**, *13*, 20160567. [CrossRef]

20. Khaled, E.M.; Atef, E. Design of novel reconfigurable frequency selective surfaces with two control techniques. *Prog. Electromagn. Res.* **2013**, *35*, 135–145.
21. Sundarsingh, S.A.B.E.F.; Ramalingam, V.S. Mechanically Reconfigurable Frequency Selective Surface for RF Shielding in Indoor Wireless Environment. *IEEE Trans. Electromagn. Compat.* **2020**, *62*, 2643–2646.
22. Wang, R.; Cai, L. Liquid Crystal Based Reconfigurable Frequency Selective Surface for ISM Applications. In Proceedings of the 2020 9th Asia-Pacific Conference on Antennas and Propagation (APCAP), Xiamen, China, 4–7 August 2020; pp. 1–2.
23. Tian, T.; Huang, X.; Cheng, K.; Liang, Y.; Hu, S.; Yao, L.; Guan, D.; Xu, Y.; Liu, P. Flexible and Reconfigurable Frequency Selective Surface with Wide Angular Stability Fabricated with Additive Manufacturing Procedure. *IEEE Antennas Wirel. Propag. Lett.* **2020**, *19*, 2428–2432. [CrossRef]
24. Engineering Simulation & 3D Design Software|ANSYS, HFSS. Available online: www.ansys.com (accessed on 29 June 2018).
25. Getting Started with HFSS: Floquet Ports, Guangzhou Yixi Technology. Available online: http://www.1cae.com (accessed on 29 June 2018).
26. Costa, F.; Monorchio, A.; Manara, G. Efficient analysis of frequency-selective surfaces by a simple equivalent-circuit model. *IEEE Antennas Propag. Mag.* **2012**, *54*, 35–48. [CrossRef]
27. Ghosh, S.; Srivastava, K.V. An equivalent circuit model of FSS-based metamaterial absorber using coupled line theory. *IEEE Antennas Wirel. Propag. Lett.* **2015**, *14*, 511–514. [CrossRef]
28. Ferreira, D.; Caldeirinha, R.F.S.; Cuiñas, I.; Fernandes, T.R. Square loop and slot frequency selective surfaces study for equivalent circuit model optimization. *IEEE Trans. Antennas Propag.* **2015**, *63*, 3947–3955. [CrossRef]
29. Joozdani, M.Z.; Amirhosseini, M.K. Equivalent circuit model for the frequency selective surface embedded in a layer With constant Conductivity. *IEEE Trans. Antennas. Propag.* **2017**, *65*, 705–712. [CrossRef]
30. Yan, M.; Wang, J.; Ma, H.; Qu, S.; Zhang, J.; Xu, C.; Zhang, A. A quad-band frequency selective surface with highly selective characteristics. *IEEE Microw. Wirel. Compon. Lett.*, **2016**, *26*, 562–564. [CrossRef]
31. Sarabandi, K.; Behdad, N. A frequency selective surface With miniaturized Elements. *IEEE Trans. Antennas. Propag.* **2007**, *55*, 1239–1245. [CrossRef]
32. Keysight Technologies, Advanced Designed System (ADS). Available online: www.keysight.com (accessed on 20 June 2019).
33. SMP1322-079LF Skyworks Diode-PIN|Skyworks. Available online: https://store.skyworksinc.com/products/detail/smp13220 79lf-skyworks/259214/ (accessed on 20 June 2019).
34. Broadband Horn Antennas. Available online: http://www.ainfoinc.com (accessed on 20 June 2019).
35. Lee, I.-G.; Park, Y.-B.; Chun, H.-J.; Kim, Y.-J.; Hong, I.-P. Design of active frequency selective surface with curved composite structures and tunable frequency response. *Hindawi Int. J. Ant. Propag.* **2017**, 1–10. [CrossRef]

MDPI

St. Alban-Anlage 66

4052 Basel

Switzerland

Tel. +41 61 683 77 34

Fax +41 61 302 89 18

www.mdpi.com

Electronics Editorial Office

E-mail: electronics@mdpi.com

www.mdpi.com/journal/electronics